Introduction to Signal Integrity:

A Laboratory Manual

Stephen C. Thierauf

Thierauf design and Consulting

Cover art and design by Chris Thierauf

Acknowledgements

Thanks to Christopher Thierauf for conceiving and designing the cover art and for physically creating the cover, and to Frank Sledd for his invaluable advice and help in navigating the self-publishing process.

0221B

Table of Contents

Preface

The most successful and creative SI engineers are those with a solid theoretical background and lots of experience correlating that theory with actual measurements and simulations.

Unfortunately, many engineers get oscilloscope measurement experience late at night on an overdue prototype or production hardware that's under close management scrutiny. This isn't the best approach. Instead, making measurements at a more leisurely pace on known good hardware under controlled conditions is a better way to learn. There are subtle (and not so subtle) differences between how a waveform should appear in theory and how it actually appears in practice. Some of this difference is caused by limitations of the oscilloscope you're using (and the way in which you use it), but the actual characteristics of the circuitry (as opposed to the ideal characteristics often assumed in textbooks) play a large part, too. To address these things this manual provides oscilloscope traces for each experiment, and Appendix C discusses oscilloscope limitations, including ways to reduce or eliminate artificial ringing caused by poor grounding techniques.

The experiments in this manual start with the basics of transmission lines and then go on to reflections and terminations. In the final chapters you'll find out about and measure the two types of crosstalk, and learn how to mitigate it. This manual provides a succinct theoretical explanation for each topic that's deep enough to guide you in making measurements. The intent is to help you bridge the gap between what you have read in a text and what you are seeing on your oscilloscope.

The Experiments

A 74HCT240 buffer/driver is used as a ring oscillator and driver to launch pulses down twisted pair wires long enough to exhibit transmission line behavior. Since the frequency of the ring oscillator can be adjusted you don't need to have access to a pulse or function generator. You only need a 5V power supply. In a pinch three or four 1.5V batteries wired in series can be used instead. The 'HCT240 is slow enough so that you don't need an expensive high-performance oscilloscope, and the measurements are somewhat forgiving of poor oscilloscope technique (a trap in which students and novice engineers or technicians sometimes get caught). The slow speed also means the circuit can be constructed on a solderless breadboard, but even so, to get the best results precautions must be taken to keep the wiring neat. Appendix A describes good wiring techniques.

As we've noted, the transmission line is a length of cable (the prototype experiments were performed on common CAT5e unshielded twisted pair, but Appendix B discusses other cables you can use).

Why Use Cable Instead of Circuit Board Traces?

The conclusions you draw from the experiments performed in this book apply equally well to circuit board traces or to wires in a cable. So, why are these experiments done over a cable instead of on a circuit board?

It comes down to cost and convenience.

Chapter 2 describes how signal integrity problems show up when the wire connecting the driver to the receiver is long enough to act as a transmission line. You'll find that the signal rise time determines if a particular length of line appears to be long. Fast signals (ones having small rise times) will experience transmission line effects even if the wire is short; slow rise times require longer wires.

The 74HCT240 used in these experiments has a slow rise time, which lets experimenters build the circuit on a solderless breadboard and to make measurements using less expensive oscilloscopes and probes. The side effect of this choice is that the transmission line must be very long before the effects we wish to measure can be easily and repeatedly seen.

A circuit board with traces can be used with these experiments, but if you choose to design a printed circuit board the traces should be no shorter than about 120" (305cm) in length. Traces shorter than this will show transmission line effects too, but a higher bandwidth oscilloscope will be required to clearly see them.

Another advantage of using cables as the transmission line is that that can be "reconfigured" more easily for various experiments than can a circuit board trace. Some of the experiments in this book examine the effects of added grounds (changes in the signal return path), which are easy to achieve when using a cable by simply connecting a wire. This can be done by using jumpers between traces, but using the cable is much simpler.

Chapter 1 Signal Integrity Background Material

What is signal integrity (SI), why should designers care about it, and when should they start thinking about it?

Signal Integrity is a specialty within electrical engineering that aims to ensure that an electrical signal reliably reaches its destination. This means the signal shouldn't have too much distortion, and it shouldn't be too affected by noise from other signals. Hidden in this description is that a signal shouldn't introduce noise strong enough to alter the proper operation of other signals or circuits.

Designers of digital logic and embedded systems care about signal integrity because ignoring it leads to hardware failures. Signal integrity problems can cause signals to fail outright, but it's more common for signals with SI problems to behave erratically (sometimes working, and sometimes not). In these situations some signals only misbehave under very specific sets of circumstances (such as a particular combination of data patterns, power supply voltages and temperatures). Clever software design (especially the use of error correction and smart recovery routines that allow the hardware to continue after an intermittent fault) can hide serious SI problems and mislead the system designer into thinking that the hardware is more robust than it actually is.

Ideally, you should consider SI during the design phase, while the schematics are being drawn and before any circuit board artwork or cabling has been created. This is often the approach taken by designers of high performance systems, and SI analysis is occasionally performed this early by very experienced designers of small or lower performance systems. These engineers know that it's not only high-speed signals that require SI attention. Under the right conditions slow signals can misbehave, too. For instance, I2C® or SPI® busses used to send data to and from a host processor can be long enough to misbehave even when they are clocked at low speeds.

In fact, signal integrity analysis is often not performed during the design phase in low performance systems, or systems that are physically small. In these systems it's more common for SI analysis to be done in the lab, on prototype or production hardware. Often this is the most expensive and time consuming way to uncover and then solve signal integrity problems.

By understanding the fundamental causes of SI problems the designer can make important choices during the design phase, often before layout has begun, that can minimize and eliminate SI problems from being designed into a piece of hardware.

The experiments and explanations in this book provide you with a basic background in signal integrity concepts and terminology, but this book is not a complete treatment. Instead, it's intended to show you how to perform experiments on actual hardware so you can see how reflections and crosstalk are created, and how you can use termination to reduce or eliminate them. You can get more information, and more detail, by consulting the books and research papers listed in the bibliography.

Terminology

SI borrows ideas and terminology from signal processing, filter theory, electromagnetics, RF and circuit theory. It also has added new terms of its own (or redefined some of the terms it's borrowed from these other disciplines). Here we explain many of the terms commonly used in signal integrity work, and that will be used throughout this book.

Wires, Traces, Transmission Lines

The copper conductor that signals travel down will generally either be the round wire in a cable or traces on a circuit board. In this book "wire" is used to mean either of these types of conductors.

A transmission line is a wire that is "electrically long". When you read Chapter 2 you'll get a better understanding of what this means, but for now we'll say that a transmission line is any wire whose resistance, capacitance and inductance appears to be evenly distributed along its entire length. Transmission lines (sometimes just called 'lines') act as storage elements, and delay signals sent down them. Pulses are free to travel up and back down transmission lines, and more than one pulse can travel along a line at the same time. If two pulses going in opposite directions pass by each other they will combine to create a new value at the place where they meet, but each will then continue to go on their separate ways.

Driver and Receiver, Near End and Far End

Signal integrity is about sending and receiving signals between integrated circuits (ICs). The ICs can be large processor chips or ASICS, or they can be small microcontrollers or other logic devices (such as 74HCT counters, inverters and so on, or the single gate 74LVC family of devices).

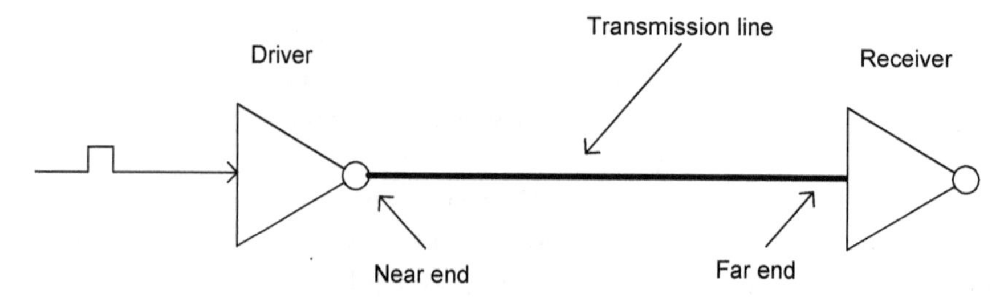

Figure 1.1 A Driver is connected at the transmission lines near end to a receiver, which is located at the lines far end.

As shown in Figure 1.1, in this book the transmitting device is called the driver. The logic device that receives the signal from the driver is the receiver. The end of the transmission line closest to the driver is called the near end. The end furthest away is the far end.

Pulse Amplitude, Rise Time and Fall Time

You'll find in Chapter 2 that it's the relationship between the signal rise time (or the fall time) and the length of the trace or wire that determines if it acts as a transmission line (and so, allows reflections to be created). The pulse rise time is defined as the time required for the signal voltage to switch from 10% of the pulse amplitude to 90% of that value. The fall time is the time the signal transitions from the 90% value to the 10% value. This "10/90 rise time" definition is shown below, in Figure 1.2.

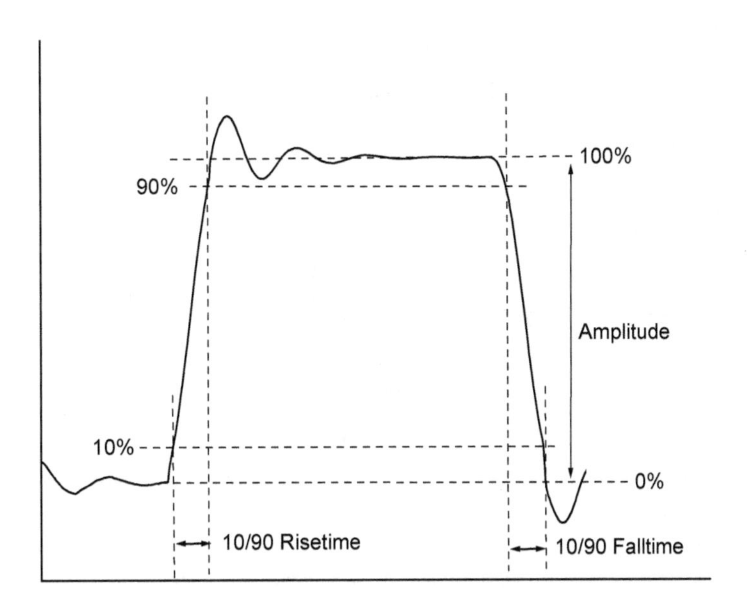

Figure 1.2 Pulse rise, and fall time, and amplitude definitions.

Rather than using the 10/90% points, some texts use the 20/80% points, and some definitions of rise time involve taking the slope of waveform as it passes through the 50% point of its swing. We'll avoid these definitions and stay with the 10/90 shown in the figure because it's widely known and is commonly used by oscilloscopes. Don't worry if your oscilloscope uses 20/80, or if the waveform you're measuring is too noisy to let you conveniently make a 10/90 measurement. The rules of thumb described in Chapter 2 are general enough so that you can successfully use any of these definitions.

Pulse Overshoot and Ringback

You'll see many times throughout this book how pulses can become badly distorted. The following figure (Figure 1.3) shows an example. The waveform exhibits overshoot (a voltage excursion that exceeds the normal, 100% level) and ringback (a pulse in the opposite direction causing the signal to dip below the 100% level). Ringback is sometimes mistakenly called "undershoot", but this is not correct; Undershoot occurs when the signal fails to rise to the 100% or 0% levels.

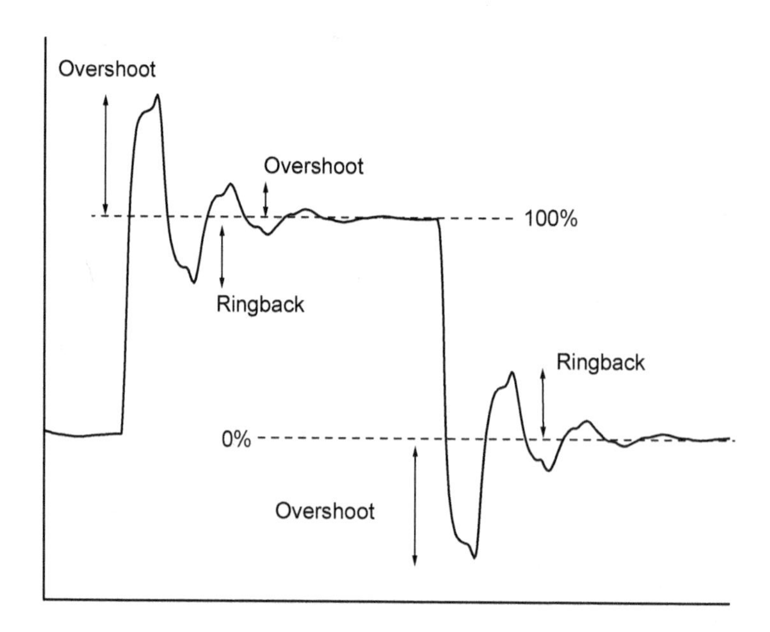

Figure 1.3 Definition of pulse ringback and overshoot.

Notice that overshoot and ringback can happen on both edges, not only the rising edge. In the falling direction we see that the signal overshoots the 0V level and takes on a negative value.

Chapter 2 Transmission Line Fundamentals

This chapter is a brief introduction to transmission lines. We discuss what a transmission line is, what it's made of, and describe its characteristics in preparation for the experiments in Chapter 3. The books listed in the bibliography go into more detail than you'll find here.

Transmission Line Effects

Figure 2.1 shows two drivers sending100ns square waves down identical 12.5" (32cm) long circuit board traces. The drivers could be the outputs from a microcontroller or ASIC, or they could be discrete logic devices located on a circuit board. The drivers have the same drive strength, but have different output rise times. Driver A's output switches from 0V to 3.3V with a *tr* = 1ns rise time, while Driver B's output rise time is10ns.

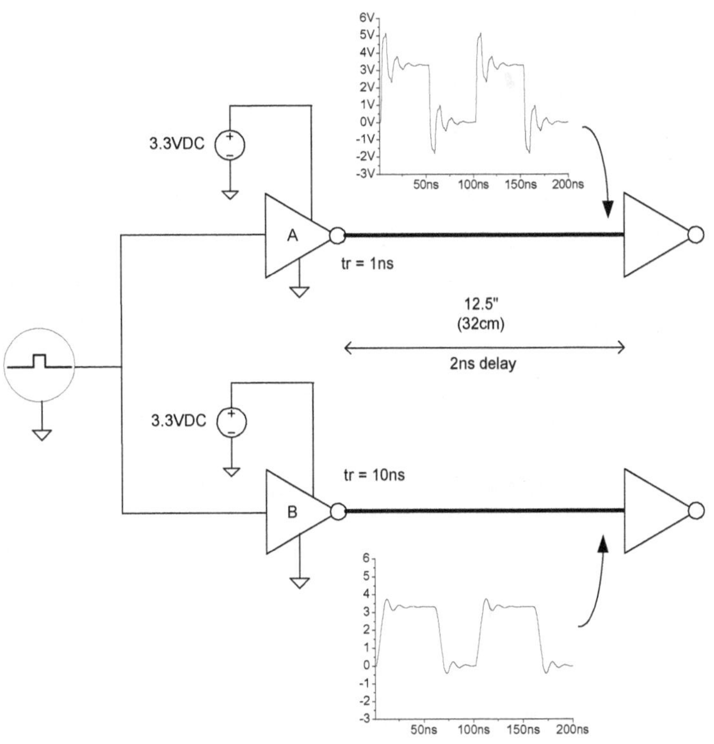

Figure 2.1 Drivers A and B send 100ns square waves down identical circuit board traces. The signal received from the 1ns rise time driver has large overshoot and ringback. The signal is greatly improved when the rise time is only 10ns.

As we see in the figure, the signals have become distorted by the time they reach their destination. Both waveforms have overshoot and ringback, but these are much worse for the 1ns rise time pulse. In fact, although the drivers are powered from 3.3V, in the 1ns rise time case the signal overshoots to more than 5V, and then rings back to 2.5V before finally settling down to 3.3V. The 10ns rise time pulse is much better behaved and shows only modest amounts of overshoot and ringback.

In the figure, the traces are long enough for the signal to require 2ns to travel from the driver to the receiver. This is important to know because the trace acts as a transmission line when its delay is long compared to the signal rise time. The trace won't behave as a transmission line when the delay is short as compared to the rise time.

Signals sent down transmission lines travel as waves and can experience attenuation (loss of signal strength) and reflections which can combine to create complex composite waveforms, including overshoots and ringback.

Signals sent down a trace or wire which is too short compared to the signal rise time to be a transmission line appear to the signal as a simple RC circuit.

This is important enough to be written as a rule (equation (2.1)), where *R* is the ratio between the delay (*td*) and the signal rise time (*tr*):

$$R = \frac{td}{tr} \tag{2.1}$$

Line lengths and signal rise times giving a value for *R* of 0.5 or larger will act as transmission lines. In this case you'll probably need to terminate the line, otherwise reflections, overshoots and ringing are likely to be present.

R values under about 0.125 indicate the delay is so small compared to the rise time that reflections, overshoots and ringback won't be created. In this situation the line appears to the driver as a capacitor and not a transmission line, so terminations aren't needed to prevent ringing.

Values of *R* between 0.125 and 0.5 are in the gray area and the reflections and overshoots may or may not be severe enough to require that you terminate the line.

We can test this rule by using it to predict the results in Figure 2.1. The *R* factor for driver A is: $R = \frac{td}{tr} = \frac{2ns}{1ns} = 2$. Since this is greater than 0.5, we'd expect the line to behave as a transmission line, and the ringing shown in the figure of the unterminated transmission line confirms this assumption to be so.

The *R* for Driver B is 0.2, which is in the gray zone where it's unclear how the line will behave. It's closer to the 0.125 "is not transmission line" limit than it is to the 0.5 "definitely is a transmission line" limit, so we'd be correct in reasoning that if any overshooting or ringback is present it should be small. And the figure shows that this is indeed the case.

Even though in the driver B case the low value for *R* properly suggested that overshoots would be small, *R* is only a figure of merit and it doesn't actually predict the severity of the overshoots. In fact, it's not correct to assume that high *R* values always mean that ringing and overshoots will be larger than for low *R* values. Other factors, such as the strength of the driver and resistance or capacitance of the load at the end of the transmission line influence the severity of the reflections. Use *R* to estimate if a given combination of driver and line will act as a transmission line or as a lumped capacitor. Don't use it to predict how bad reflections or ringing are likely to be.

To summarize:

It's not the frequency of the pulse that determines if a signal will experience reflections, overshooting and ringback. What matters is the rise time of the signal as compared to the amount of time the signal takes to reach its destination.

This explains why high frequency signals are particularly likely to have signal integrity problems: high frequency signals must have a small rise times, which means short lengths of wire or traces can be long enough to have large *R* values (and so, act as transmission lines). In chapter 5 you'll see how the pulse width and frequency affect the pulse shape and overall appearance.

Resistance, Inductance and Capacitance

If we were to connect an ohmmeter to the ends of a 5 mil[1] (0.127mm) wide circuit board trace 20 inches (51cm) long we'd find it has a resistance of about 2Ω. The actual value depends on how thick the trace is, and other manufacturing details, but 2Ω is a good approximation for a 5 mil wide copper trace manufactured on common circuit boards. Although the ohmmeter doesn't show it, there is also inductance and capacitance present. Depending on the circuit boards construction the inductance would be between about 130 to 200nH; the capacitance would range from about 50 to 85pF.

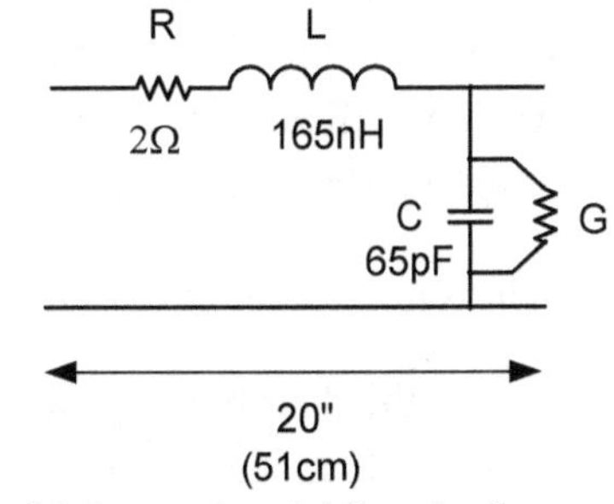

Figure 2.2 Lumped model for a 5 mil (0.127mm) wide circuit board trace the total resistance, inductance capacitance and conductance.

As you can see in Figure 2.2, the inductance is in series with the resistance, and the capacitance is present from the trace to other traces, or a ground or power plane in the stackup. A dielectric material (such as the insulation coating a wire, or the FR4 circuit board material) separates the signal trace from the *signal return path*. The electrical characteristics of the insulator (in particular, the dielectric constant, which we'll abbreviate here as *Dk*) is an important factor in determining the actual capacitance.

Resistor *G* is placed in parallel with the capacitor *C* to account for its leakage, but for common FR4 type circuit boards *G* has such a large value (the leakage is so small) that it can be ignored unless we are making the measurements in the GHz range. We won't need to know about *G* for any of the examples in this book.

[1] 1mil = 0.001 inches, so 5 mils = 0.005 inches

G= leakage resistance

In the figure resistor R is the resistance of trace as measured at DC with an ohmmeter. The resistor converts some of the signal energy flowing through it into heat, and so acts as a loss. At high frequency the resistance is greater than the DC value. For instance, for traces on the board's surface, the resistance at 1GHz will be about 20Ω (and it'll climb to nearly 45Ω at 5GHz). This change in resistance with frequency is created by the *skin effect* and causes more loss at high frequencies than at low ones. This is a major reason signals traveling down wires or traces are distorted since each of the harmonics making up a pulse experiences a different amount of resistance (and so, loss). When these misadjusted harmonics recombine at the far end they create a pulse having a very different shape than it did when it was launched at the near end.

Inductor L is the amount of inductance present between the wire and the path the signal takes to get back to the driver (the signal return path). The amount of inductance depends on how far away the signal wire or trace is to the return path, but in the schematic this isn't as obvious as with capacitance. In the figure the capacitor clearly shows a connection between the signal wire and the return, but the inductor is just in series with the resistor without any direct connection to the return (which is usually ground). From this we might think the inductance is a property of the wire or trace much like resistance. But, although the symbol doesn't imply it, the inductance gets larger the further away the signal is away from the return path. This is just the opposite from capacitance, which gets smaller as the distance increases.

Distributed Parameters

The total amount of resistance, capacitance and inductance shown in the figure is distributed smoothly all along the length of the trace and doesn't appear as a single lump the way the figure suggests. Because of this we can divide the values in the figure by the length of the trace to get the values *per unit length*. For instance, the inductance has a value of $\frac{165nH}{20''} = 8.25nH / inch$. In theory, the distributed parameters are infinitesimally small, but in practice we only need to make them small enough so they appear that way in our particular situation. This is shown below, in Figure 2.3, where the 20" long line has been divided into 20,000 segments, each 1mil in length.

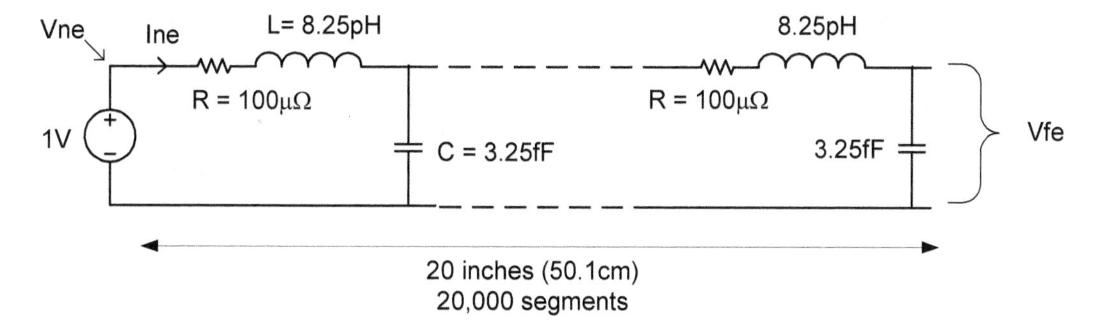

Figure 2.3 Distributed Model for a 20 inch long circuit board trace divided into 20,000 segments, each representing 0.001 inch in length. The signal is launched at the near end (Vne) and is received at the far end (Vfe).

Voltage and Current

If you were to instantly connect the 1V battery shown in the figure to the transmission lines near end and immediately measure the current you might expect from Ohm's law the current to be $Ine = \frac{V}{2} = \frac{1}{2} = 500mA$, since the line has a 2Ω DC resistance. In fact, in this situation you'd find the current to be 50 times less than the voltage: $Ine = \frac{V}{50} = \frac{1}{50} = 20mA$.

You'd also find a delay of 3.3ns had to elapse before the voltage appeared at the traces far end (*Vfe*). When the voltage did arrive you'd discover it wasn't the 1V you were expecting, but instead had become 2V. If you waited long enough (say, 10 or 20 times the 3.3ns delay of the line) you'd find that the current had dropped to zero and the far end voltage (*Vfe*) had become the 1V you had been expecting in the first place.

At least momentarily the relationship between the voltage and current followed an Ohms law behavior: As far as the battery is concerned, the trace has initially acted like a 50Ω resistor. Given enough time the trace in this example acts as an open circuit and draws no current from the battery. Early telegraph operators and electrical engineers observed this when signaling over very long distance wires, and called this phenomenon the telegraph lines "surge impedance". Signal integrity engineers simply call it the impedance.

Impedance

Circuit analysis can be used to find the impedance and delay of a long chain of RLC lumps such as the ones shown in Figure 2.3.

The circuit analysis is made much easier if we assume the effects of losses (*R* and *G*) are small compared to the effects of the inductance and capacitance (*L* and *C*). This is a good assumption for most circuit board traces when the frequency is greater than a few MHz. At lower frequencies the resistance dominates the inductance and the impedance is a complex number (one having real and imaginary parts). At high frequency the inductances influence is greater than the resistance. In this case the impedance is a real number (one that doesn't have an imaginary part).

Provided this is so the *characteristic impedance* Z_O of the trace can be found with equation (2.2):

$$Z_O = \sqrt{\frac{L}{C}} \tag{2.2}$$

This is called the characteristic impedance because it describes a fundamental characteristic of the transmission line. In fact, it's properly called the *lossless characteristic impedance* because we've assumed the losses are small enough to ignore. The impedance of lossless transmission lines appears as a simple resistance and doesn't change with frequency (it's the same for every frequency).

Effects of Crosstalk

Strictly speaking, Equation (2.2) only predicts the correct value for Z_O when the trace is far enough away from other transmission lines so that crosstalk and other coupling can't occur. The impedance will have a different value (which could be higher or lower than Z_O) depending on how close neighboring traces are, and the data pattern in which they switch. To distinguish the impedance measured under these kinds of switching conditions from the characteristic impedance it's given a different name (such as the "even mode" or "odd mode" impedances, abbreviated Z_{OE} or Z_{OO}). This difference in definition is subtle and is often overlooked, but it's an important factor in understanding signal noise, especially when traces are routed close to each other, or when wires in a bundle are packed closely together. The experiments in the next chapter show this by having you measure the cable impedance when other signals in the bundle switch, or when they are connected to ground.

To summarize:

> **The impedance of a line with no losses (or losses low enough that they can be ignored in your application) appears as a resistance that has the same value at all frequencies, provided the line doesn't experience coupling from other lines. The value is equal to** $Z_O = \sqrt{\frac{L}{C}}$.

If the losses are too high to ignore the impedance becomes a complex number (one that has real and imaginary parts that represent a magnitude and the phase angle between the voltage and the current). The impedance of lossy lines can't be represented by a simple resistor and it changes with frequency. Lossy transmission lines aren't explored in this book, but the bibliography lists references where they are discussed.

Example Impedance Calculation

The lossless characteristic impedance of one of the small lumps in Figure 2.3 is:

$$Z_O = \sqrt{\frac{L}{C}} = \sqrt{\frac{8.25pH}{3.25fF}} = 50\Omega$$

We get the same characteristic impedance when the values from the lumped model of Figure 2.2 are used:

$$Z_O = \sqrt{\frac{L}{C}} = \sqrt{\frac{165nH}{65pF}} = 50\Omega$$

From this we see that (unlike resistance) impedance is independent of length.

Why the Impedance Doesn't Change Even When the Length Does

The impedance doesn't change because as equation (2.2) shows, the impedance equation uses the ratio of inductance and capacitance, and (as we saw in the above example) for a uniform line (one whose inductance and capacitance doesn't change as the line gets longer) the ratio remains the same no matter how long or short we make the line.

Delay

Circuit analysis of the distributed circuit in Figure 2.3 can be used to find by how much a trace delays a signal. Provided we can ignore losses (R and G) as we did for the impedance, the lossless time delay td from one end of the trace to the other is found with equation (2.3):

$$td = \sqrt{L \times C} \qquad (2.3)$$

Example Delay Calculation

For example, to calculate the delay per segment of the circuit board trace in Figure 2.3 we find:

$$td = \sqrt{L \times C} = \sqrt{8.25pH \times 3.25fF} = 160 fs/mil$$

Since there are 1000 mils per inch, this is the same as 160ps/inch.

Frequency doesn't show up in the equation, which tells us that a lossless transmission line delays all signals equally. A 10MHz signal sent down this transmission line will experience the same amount of delay time as will a 100MHz signal: Each will be delayed by 160ps for every inch of length.

It's important to note that lossy transmission lines don't behave in this way: With those lines the delay does depend on frequency. This can cause signals that are made from more than one frequency (such as digital pulses, or audio signals) to become distorted as each frequency component (harmonic) is delayed by different amounts.

Why the Delay Changes When the Line Gets Longer

We've seen how the impedance of a transmission line is the same no matter how long the line, and we've just seen that delay increases with length. This makes senses: We'd expect a longer linc to delay a signal more than a shorter one. But why?

We already know that a longer trace or wire will have more inductance and capacitance than a short one. We notice from equation (2.3) how the delay involves the product of the inductance and capacitance rather than their ratio, so longer lines (which naturally have larger values of inductance and capacitance) result in longer delays. But the delay increases as the square root of the product, so the delay remains practically constant for small changes in inductance or capacitance. This means you don't have to precisely measure L and C to get a good estimate of the delay.

Some Typical Delay and Parameter Values

Table 2.1 lists typical impedance, delay, inductance and capacitance values for some common types of transmission lines.

Table 2.1 Typical characteristics of circuit board traces and common cables.

Type	Delay ps/inch (ps/cm)	Impedance Ω	Capacitance pF/inch (pF/cm)	Inductance nh/inch (nh/cm)
Air	85(33)			
RG8 coax	98 (39)	50	1.96 (0.77)	4.9 (1.93)
RG62A coax	101 (40)	93	1.1 (0.43)	9.4 (3.7)
RG6 coax	110 (43)	75	1.47 (0.58)	8.25 (3.25)
CAT5e cable	120 (47)	100	1.2 (0.47)	12 (4.72)
RG316 coax	122 (48)	50	2.44 (0.96)	6.1 (2.4)
FR4 microstrip with soldermask	160 (63)	30 - 200Ω	5.3 – 0.8 (2.1- 0.3)	4.8 – 32 (1.9 – 12.6)
FR4 stripline	180 (71)	20 - 100Ω	9 – 1.8 (3.5 – 0.7)	3.6 – 18 (1.4 – 7.1)

From the table we see that the capacitance and inductance (and so, the impedance and delay) depends on the type of transmission line.

For instance, miniature RG316 coaxial cable has an impedance of 50Ω, and will delay signals by 122ps for every inch of its length (which is the same as 48ps per cm). It has a capacitance of 2.44pF per inch (0.96pF per cm), and an inductance of 6.1nH per inch (2.4nH per cm). RG8 coax has the same impedance but is "faster" (98 ps per inch).The CAT5e cable used in the experiments has a higher impedance (100Ω) and typically delays signals by 120ps per inch (47ps/cm).

All of these cables are faster than circuit board traces on FR4 (the material often used to create common circuit boards). Traces on the surface (microstrip traces) typically delay signals by 160 ps for every inch of length (63ps/cm), while traces embedded in the board and between ground planes introduce a delay of around 180ps/inch (71ps/cm).

Using the Voltage Divider Principal to Determine Impedance

Review the series circuit shown in Figure 2.4. At first glance this circuit seems too simple to spend much time on, but we'll use it to learn how to find the characteristic impedance of transmission lines from measurements made with an oscilloscope. It will lead us to an understanding the TDR (Time Domain Reflectometer), an instrument commonly used in signal integrity.

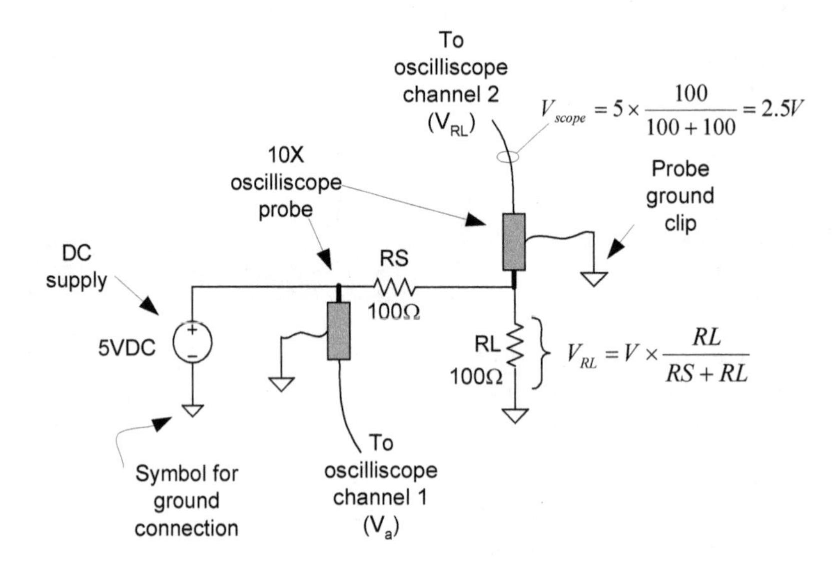

Figure 2.4 Voltage divider circuit. Oscilloscope channel 1 measures V_a; channel 2 measures V_{RL}.

We know that the current flowing through resistor *RL* determines the voltage across it. But by using circuit theory we can take a shortcut to find the voltage across *RL* without first calculating its' current. We'll find this approach to be very helpful when we turn our attention to transmission lines. In fact, we'll use this method to determine the impedance of the cable we'll be using as a transmission line. You can do the same thing with a higher bandwidth oscilloscope to find the impedance of traces on a circuit board.

The voltage appearing across *RL* depends on the voltage applied to resistor *RS* and is in proportion to *RL*'s resistance as compared to the total resistance in the circuit. This is the *voltage divider principle*, and is shown in equation (2.4).

$$V_{RL} = V_a \times \frac{RL}{RS + RL} \tag{2.4}$$

By using the equation we see that the voltage across load resistor *RL* is equal to the applied voltage (V_a) times the ratio of *RL* to the total resistance (*RL* plus the series resistor *RS*). The worked example in the figure shows how the voltage will be divided by two when *RS* and *RL* are equal. Since oscilloscope channel 1 shows that 5V is being applied to the circuit, channel 2 displays 2.5V.

We have two more things to learn from the simple voltage divider.

The first involves the connection of the oscilloscope. From the figure we see that the ground clip of each oscilloscope probe is attached to ground, which is the common connection between the power supply and the bottom terminal of resistor *RL*. In fact, channel 2 is not just measuring the node voltage between *RS* and *RL*. The deeper insight is that it's displaying the voltage across *RL*.

The second observation is that if *RL* were to be replaced with a black box we'd be correct in saying that because the box has 2.5V across its inputs (one of which is ground) it has an input resistance *with respect to ground* of 100Ω. In effect, we've determined the boxes' input resistance (*RL*) by knowing the value of *RS* and measuring the voltage across the box. We've done this without knowing what's actually inside the box. We'll come back to this important observation shortly when we measure transmission line impedance.

By solving equation (2.4) for *RL* we'll have a powerful tool that allows us to determine the input resistance of a box if we know the resistance of *RS* and the voltage V_a. This has been done in equation (2.5).

$$RL = \frac{V_{RL} \times RS}{V_a - V_{RL}} \tag{2.5}$$

This method gives accurate results because it depends on measuring the difference between two voltages rather than precisely measuring their actual values. Because of this we can find the correct value for *RL* with reasonable accuracy even when using a poorly calibrated oscilloscope. The only requirement is that we know the true value of resistor *RS*. This is why it's suggested in the experiments to measure the resistance with an accurate digital multi-meter or a resistance bridge before performing the experiments.

We can test this equation by returning to the example shown in the previous figure. The setup is shown in Figure 2.5.

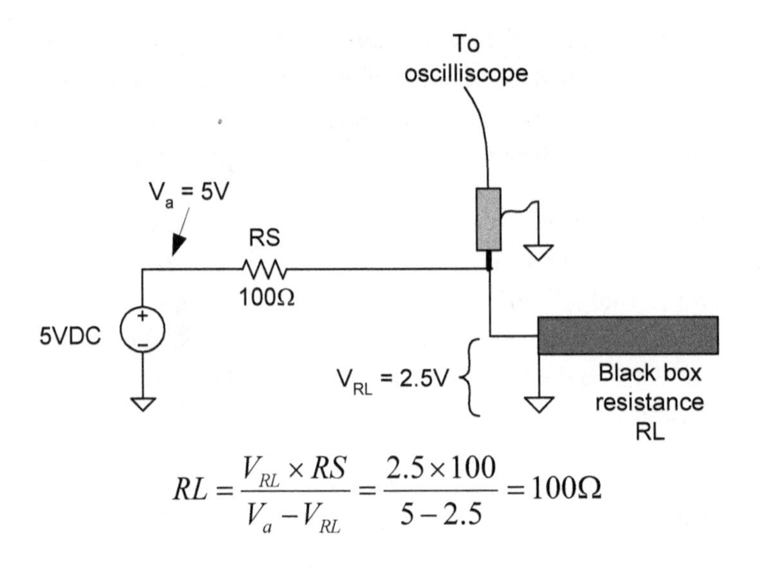

$$RL = \frac{V_{RL} \times RS}{V_a - V_{RL}} = \frac{2.5 \times 100}{5 - 2.5} = 100\Omega$$

Figure 2.5 Using the voltage divider principal to find an unknown resistance.

The voltage across resistor *RL* (V_{RL}) is 2.5V, and the output voltage from the power supply V_{out} is 5V. Series resistor *RS* has a value of 100Ω.

As shown in the figure, we find *RL* to be 100Ω by using equation (2.5). This matches the results we already know from the previous discussion. From this we conclude we can use equation (2.5) to determine the input resistance of a black box connected to a known resistance and voltage. In the next chapter we'll see how by applying a pulse instead of a DC voltage this technique can be used to find the input impedance of a transmission line.

Chapter 3 Laboratory Exercises: Impedance and Delay

In this chapter you'll measure impedance and the delay of a transmission line. As you saw in the previous chapter, the impedance can be found by applying the voltage divider principal and we'll use this idea to estimate the impedance of your cable. You'll also explore how the switching activity of adjacent wires can cause a transmission lines impedance to change (this is called the line's "odd" and "even modes"), and we'll see how improving the signal return path by grounding some of the previously unused wires in the cable alters the impedance.

Experiment 3.1: Voltage Divider Oscilloscope Measurement

Purpose

To learn how an oscilloscope and pulse generator can be used to determine the value of an unknown resistance.

Procedure

1. The test setup is shown in Figure 3.1. Be sure to check the build notes in Appendix A for construction details and circuit operation, and for information regarding parts selection.

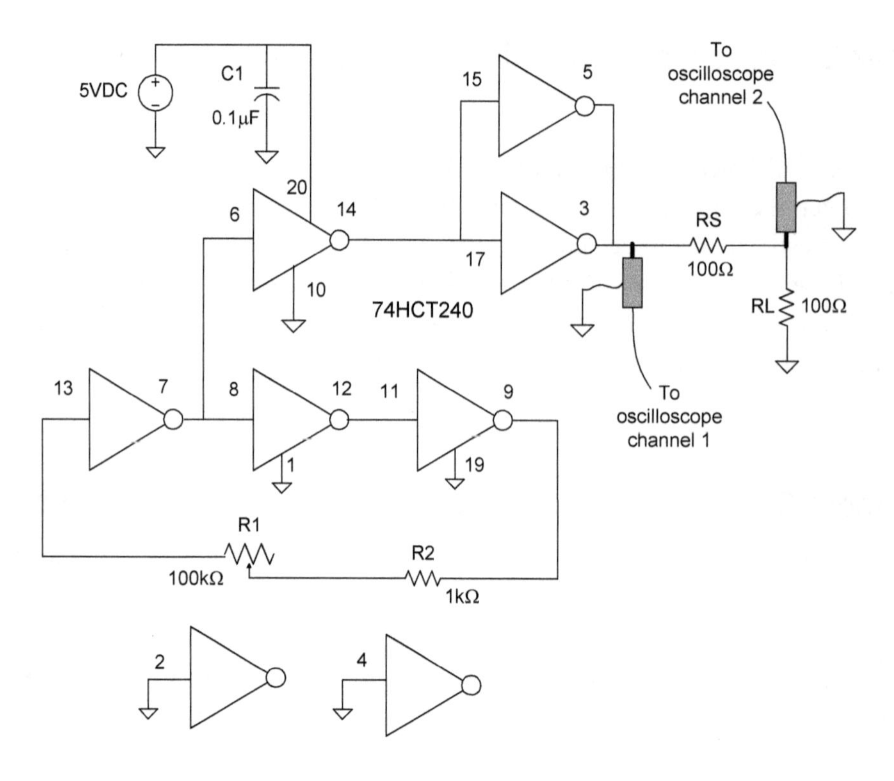

Figure 3.1 Hardware setup for Experiment 3.1.

2. Use a meter with an ohms function to find the actual value of *RS* and *RL* before you connect them to the circuit. A digital multimeter (DMM) is preferred. Record their values in your notebook, then connect them as the figure shows. We'll be comparing the value of *RL* found with your meter in this step with the value you'll determine in the following steps. In the prototype $RS = 98.2\Omega$ and $RL = 98.3\Omega$.

3. Connect the oscilloscope probes as shown in the figure.
   a. Set the trigger to occur on the rising edge of channel 1.
   b. You'll get cleaner, more easily measured waveforms when the oscilloscope probe's ground connection is made directly to pin 10 of the integrated circuit, and the probe ground lead is kept short. See Appendix C for details.

4. Apply power to the circuit and adjust resistor *R1* so that the low to high going pulse appearing on channel 2 is at least 200ns wide. Figure 3.2 shows an example of the prototypes 330ns wide pulse used in this experiment. The actual width isn't critical, but pulses wider than 200ns are convenient when using this setup in the following experiments.

5. Apply power to the circuit and record in your notebook the voltages observed on the oscilloscope. Figure 3.2 shows the voltages measured in the prototype.

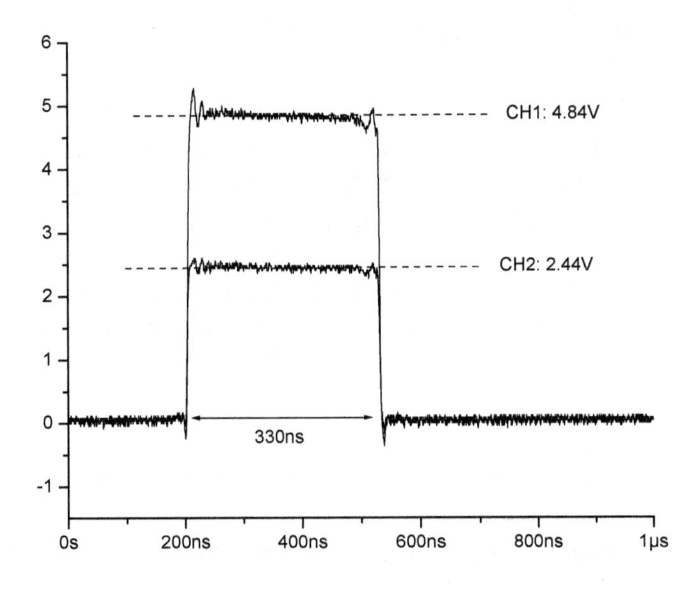

Figure 3.2 Oscilloscope waveforms from Experiment 3.1.

6. Use equation (2.5) from Chapter 2 to determine the value for *RL*, and compare it to the actual value determined in step 2. Note that in this setup V_a is the output voltage from the integrated circuit, which is measured with oscilloscope channel 1. V_{RL} is measured on channel 2.

Example Calculation

The following values were measured on a test setup using the procedure described above:

RS $= 98.2\Omega$ (value from step 2)
RL $= 98.3\Omega$ (measured value from step 2, and used to compare against calculated value)
V_a $= 4.84\text{V}$ (value from step 4, on oscilloscope channel 1)
V_{RL} $= 2.44\text{V}$ (value from step 4, on oscilloscope channel 2)

$RL = \dfrac{V_{RL} \times RS}{V_a - V_{RL}} = \dfrac{2.44 \times 98.2}{4.84 - 2.44} = 99.8\Omega$. This is within 1.5% of the known value of 98.3Ω determined in step 2.

Comments

Notice that the waveforms are not the ideal you'd obtain from simulation. Even with this simple circuit the oscilloscope displays some overshoot, and the waveform tops are not perfectly flat. The downward slope and thickness of the trace on the screen makes it difficult in practice to precisely measure the voltage. This uncertainty can result in large computational errors. For instance, if we had decided that V_a was 4.80V rather than 4.84V we would have calculated a value of 102Ω for *RL*. A 40mV change in the reading would more than double the error from 1.5% to greater than 3%.

From this experiment we conclude that the value of a resistor can be determined to reasonable accuracy by using an oscilloscope to observe the pulse response, but to insure accuracy the voltages must be measured with care.

Supplemental Exercises

- Replicate this experiment with other values for *RL* so that you become proficient in measuring the voltages accurately.
- How does the voltage on channel 1 change with different values for *RL*? Why does it behave in this way?
- Replace *RS* with various values of resistors between 50Ω to1kΩ and repeat this experiment with *RL* = 100Ω. What is the optimum relationship between *RS* and *RL*?
- Replicate this experiment but use long ground wires on the oscilloscope probes. Try other points where ground is present on the board rather than attaching them to pin 10 of the integrated circuit. Move the oscilloscope ground connections to various locations and see how doing so affects the waveforms. Why does this happen? Become proficient in using the techniques described in Appendix C to improve the waveform.

Experiment 3.2: Measuring Characteristic Impedance of a Transmission Line

Purpose

To measure the characteristic impedance of a transmission line by applying the techniques developed in Experiment 3.1.

Procedure

1. The test setup is shown in Figure 3.3. It's identical to the setup used in Experiment 3.1, except resistor *RL* has been replaced with a length of cable. If you have not already performed Experiment 3.1 be sure to check the build notes appearing in Appendix A for construction details and circuit operation, and for information regarding parts substitution.

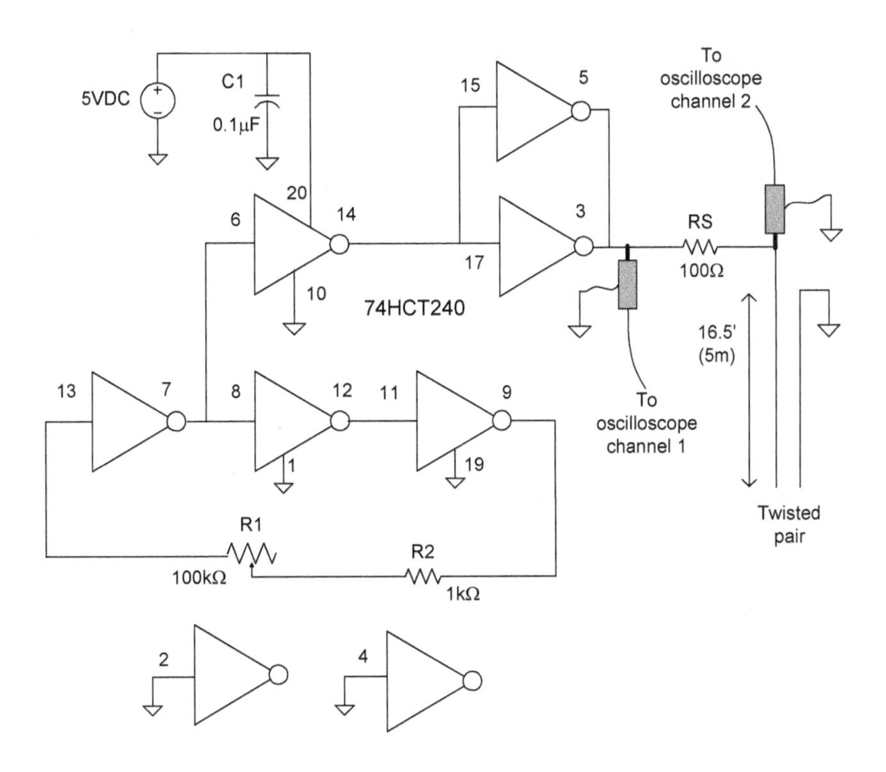

Figure 3.3 Hardware test setup for Experiment 3.2.

2. Use a meter with an ohms function to find the actual value of *RS* before you connect it to the circuit. A digital multimeter is preferred. Record its value in your notebook. In the prototype *RS* had a value of 98.2Ω.

3. Prepare 16.5 feet (5 meters) of CAT5e cable as described in Appendix B. Connect the white wire with the blue stripe to ground on the circuit. Don't yet connect the wire with the solid blue color to resistor *RS*. It will be connected later, in step 7.
    a. Note: As shown in the figure, only one end of the wire is connected to the circuit. The other end (the far end) is unconnected. For proper operation of this experiment it's critical to verify that the ends are not touching each other, or other metal objects.

4. Connect the oscilloscope probes as shown in the figure.
    a. Set the trigger to occur on the rising edge of channel 1.
    b. You'll get cleaner, more easily measured waveforms when the oscilloscope probe's ground connection is made directly to pin 10 of the integrated circuit, and the probe ground lead is kept short. See Appendix C for details.

5. Apply power to the circuit and adjust resistor R1 so that the pulse appearing on channel 1 is at least 200ns wide. The actual width isn't critical, and provided the width is more than 200ns there is no need to change this setting if you've already set this up in Experiment 3.1.

6. Observe the waveform on channel 2. It should be essentially identical to the waveform shown on channel 1.
    a. There is a wiring error or a problem in the oscilloscope settings, or the way in which it's connected to the circuit if the two waveforms do not have the same amplitude, rise time, width and nearly identical ringing. Be sure the two probes are grounded to the same point, and keep the grounds short. As part of your trouble shooting verify that the two oscilloscope probes are working properly. Replace them if necessary. Consult Appendix C for guidance and techniques for reducing ringing. Do not proceed to the next step in this experiment until the two waveforms are nearly indistinguishable.

7. Connect the solid blue colored wire to resistor *RS* and notice that both the channel 1 and 2 waveforms have changed.
    a. Adjust the oscilloscope time base and horizontal position controls to zoom in on the plateau (the flat portion) appearing on the rising edge of the channel 2 waveform. Figure 3.4 shows an example. The insert in the figure shows how the

entire pulse displayed on channel 2 looks before zooming in on just the rising edge. The plateau in the figure is 48nS wide and occurs at 2.52V. The voltage and width you measure in your setup will differ, depending on the type and length of cable you're using, the actual value of *RS*, and on the precise electrical characteristics of your particular 74HCT240 integrated circuit.

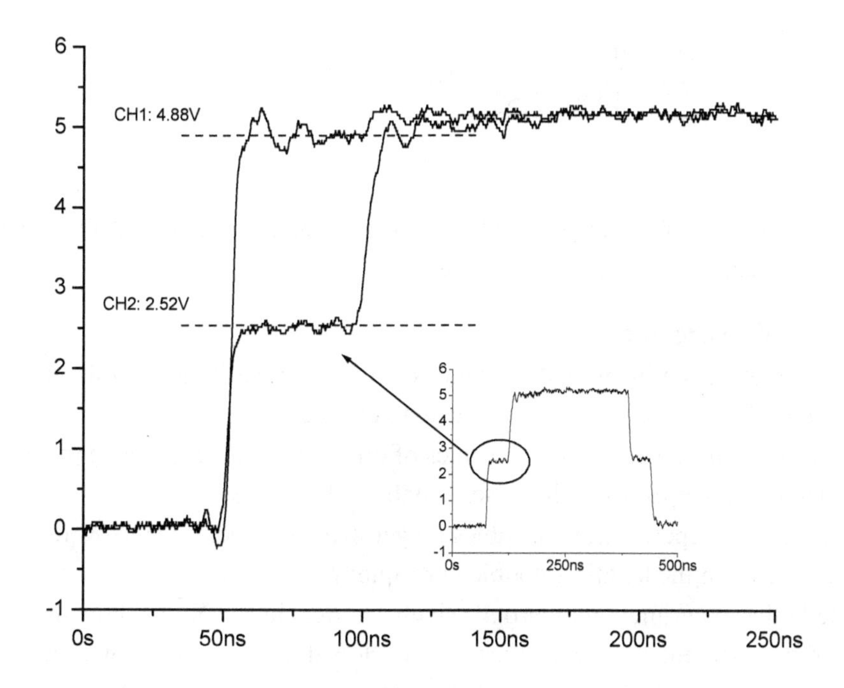

Figure 3.4 Example measured results for Experiment 3.2.

8. Use equation (2.5) from Chapter 2 to find the input resistance of the transmission line. Record the value in your notebook.

Example Calculation

The following values were measured on a test setup using the procedure described above:

RS = 98.2Ω (value from step 2)
V_a = 4.88V (value from step 7 and Figure 3.4)
V_{RL} = 2.52V (value from step 7 and Figure 3.4)

$$RL = \frac{V_{RL} \times RS}{V_{out} - V_{RL}} = \frac{2.52 \times 98.2}{4.88 - 2.52} = 105\Omega .$$

Comments

The actual impedance of this cable is 100Ω as determined by a dedicated TDR instrument; the calculated impedance of 105Ω is 5% too high.

Supplemental Exercises

- Repeat this experiment with *RS* values ranging from 22Ω to 1kΩ. Notice how the various values of *RS* cause the plateau voltage to change.
- Try this same experiment with cables of different types, and notice the difference in the impedance. Appendix B discusses suitable cable types.
- Repeat this experiment with cables longer than 16.5 feet (5m). How does the waveform change when the length is doubled or tripled?
- Redo this experiment with progressively shorter cables. Notice how the plateau changes, and the apparent change in the rise time. How does the waveform look when the cable is 2 feet (60cm) long? How does it look when the cable is 1 foot (30cm) long?
- Lay a single 16.5 foot (5m) long insulated single conductor wire on the floor. The wire can be any diameter, but AWG #20 to #26 sizes are easy to work with. The wire type may be stranded or solid. Use tape to secure it to the floor, and keep the wire straight (do not bunch the wire into a coil). Place an identical wire immediately adjacent to the first and tape it to the floor alongside the first wire. It's desirable for them to be touching along their entire length. Disconnect all other wires from resistor *RS* and attach one of the floor wires to the circuit board ground, and the other wire to *RS*. Use the test setup to find the impedance. Separate the wires by 0.5 inches (1.3 cm) and measure the impedance again; try other separations and record their impedances. Create a plot of impedance vs. distance.

Experiment 3.3: Effects of Other Return Paths on Impedance

Purpose

To observe how transmission line impedance changes when additional wires in a cable are connected to ground.

- Note: The CAT5e cable used in these experiments has a total of 8 wires. These are grouped into 4 two wire pairs (twisted pairs), as described in Appendix B. This experiment will use one of the twisted pairs (the blue pair), and two other wires in the cable. The color of the wires used isn't important and won't alter the results in these experiments[2], but for clarity and to be consistent we'll call out wire of specific colors. The assumption in this experiment is that the blue pair was used in the previous experiment as the signal wire, and the orange and brown wires will be connected as described next. The wires are prepared at one end (the near end) only. The far end isn't connected.

Procedure

1. The test setup is shown in Figure 3.5. It's identical to the setup used in Experiment 3.2, except additional wires within the cable will be connected to ground. If you have not already performed Experiment 3.2 be sure to check the build notes appearing in Appendix A for construction details and circuit operation, and for information regarding parts substitution.

[2] In fact, the different colored pairs are electrically similar, but they are intentionally slightly different by design. The differences are too small to be detected by the equipment we're using here. For our purposes in these experiments we're right to assume the pairs are electrically identical.

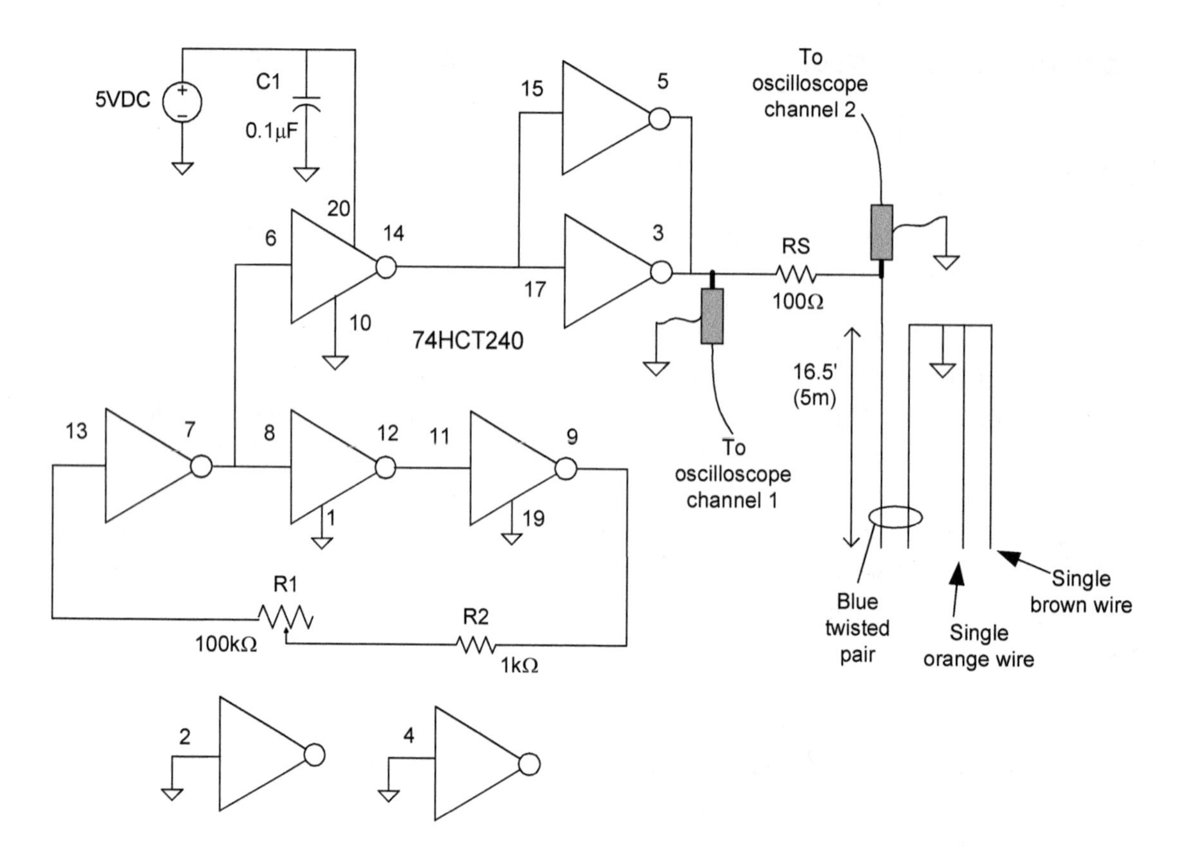

Figure 3.5 Setup for Experiment 3.3.

2. Prepare the cable as described in experiment 3.2. Then untwist the orange pair so they become two separate wires. One of the wires will be all orange; the other will be white with an orange stripe. Now untwist the brown pair. The length where the wires are untwisted isn't critical but it shouldn't exceed 5 inches (13cm). Remove about ¼ inch (0.6cm) of insulation from the orange wire, and then from the brown wire. The white wire with the orange stripe and the white wire with the brown stripe are not used and won't be connected. Four wires (the blue pair, and the orange wire and the brown wire) are now ready to connect to the circuit -- see Figure 3.6

3. In this step connect only the blue twisted pair as was done in Experiment 3.2.

4. Repeat Experiment 3.2 and calculate the impedance to get a base line measurement with which the following measurements will be compared. Identify this as the "1 gnd" case.
    - Tip: To obtain the cleanest waveforms set the oscilloscope to trigger on the rising edge of oscilloscope channel 1

5. Connect the orange wire to ground. Leave the brown wire unconnected. The cable now contains 2 grounded wires and one signal wire. Measure the plateau voltage displayed on channel 2 and then calculate and record the impedance as was done in step 4. Identify this as the "2 gnd" case.

6. Connect the brown wire to ground; leave the orange wire still connected to ground. The cable now contains 3 grounded wires and one signal wire. Measure and record the impedance as you did in steps 4 and 5. Identify this as the "3 gnd" case. This connection is shown below, in Figure 3.6.

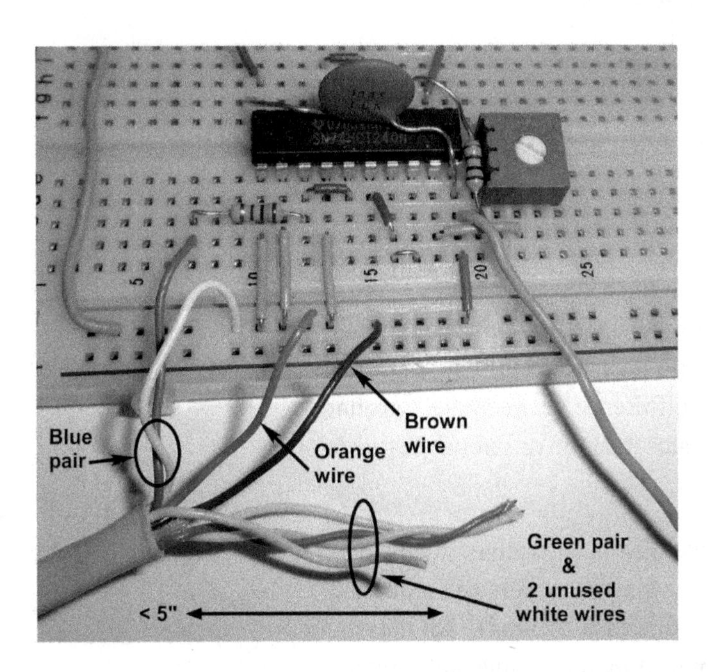

Figure 3.6 Wiring when 3 wires are connected to ground. The remaining 4 wires are unused.

7. Figure 3.7 shows an example measurement and the resulting impedance calculations where the results from steps 4 – 6 are overlaid. The impedance has changed as shown in the figure.

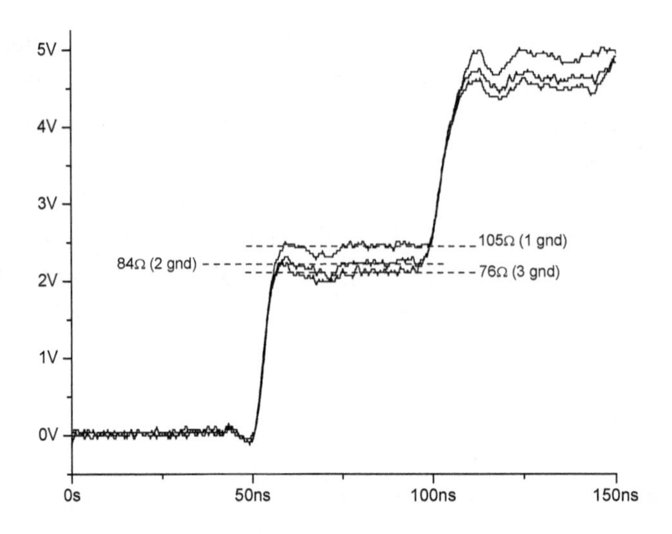

Figure 3.7 Changes in impedance when more ground wires are added.

Comments

By connecting wires within the cable to ground we've changed the impedance of the signal wire. In fact, the impedance falls as more grounds are connected, but the change is not linear: The difference between the 1 ground and 2 ground cases is greater than the difference between the 2 and 3ground cases.

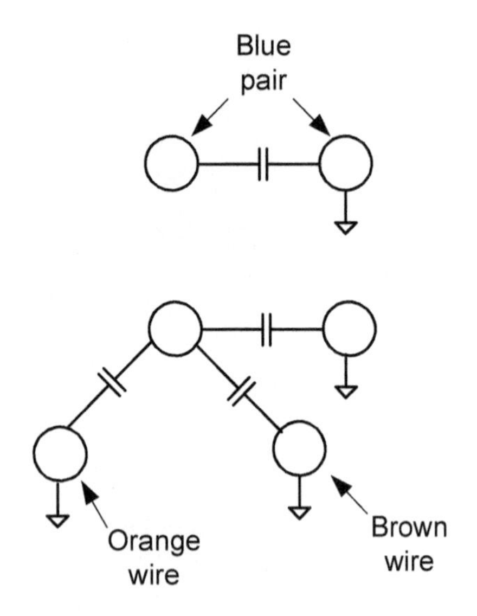

Figure 3.8 Connecting wires to ground increases capacitance and so decreases its impedance.

The impedance of the blue wire is determined by the amount of inductance and capacitance between it and any wire acting as a return (in this case, ground). An important observation is that by connecting the orange and brown wires to ground we've added capacitance to the blue wire, but its inductance remains unchanged. From Chapter 2 we know the impedance will fall as the capacitance increases. Therefor we would expect the impedance to be lower as more grounds are added because by doing so we've increased the capacitance present on the signal but haven't added an equal amount of inductance, and we see this effect in this experiment. Figure 3.8 shows how the capacitance increases because the 3 capacitors are in parallel.

This phenomenon doesn't only occur with wires in a cable. Circuit board traces behave in this way, too.

Supplemental Exercises

- Repeat this experiment but connect the orange and brown wires to +5V rather than ground. Does the impedance change?
- Connect the blue wire to *RS* and disconnect all other wires in the cable. How does the waveform change? Why?
- Connect the blue wire to *RS* and only connect the brown wire to ground. Disconnect all other wires in the cable (including the white wire with the blue stripe). Measure the impedance. Why is it different from the impedance obtained when the white wire with the blue stripe is used as a ground?
- Connect the remaining wires in the cable to ground, one at a time, for a total of 7 ground wires. Plot the impedance vs. number of grounds. What is the trend? How does the impedance differ from the case where only 3 wires are connected to ground?

Experiment 3.4: Switching Modes

Purpose

To observe how transmission line impedance changes when one other wire in the cable switches at the same time as the signal carrying wire. This demonstrates even and odd mode switching behavior.

Procedure

1. The test setup is shown in Figure 3.9. It's similar to the setup used in Experiment 3.2, except in this experiment two rather than only one wire will be caused to switch. If you have not already performed Experiment 3.2, then before proceeding with this experiment check the build notes appearing in Appendix A for construction details and circuit operation, and for information regarding parts substitution.

2. Prepare the cable as described in Experiment 3.2. If you've not already done so as part of Experiment 3.3, untwist the orange pair at one end of the cable so they become two separate wires. One of the wires will be colored all orange; the other will be colored white with an orange stripe. The brown pair will not be used in this experiment. The length where the wires are untwisted isn't critical but it shouldn't exceed 5 inches (13cm). Remove about ¼ inch (0.6cm) of insulation from the orange wire. The white wire with the orange stripe isn't used and won't be connected. Three wires (the blue pair, and the orange wire) are now ready to be connected to the circuit.

3. Record and measure the value of *RS*. You already have this value if you've previously performed experiment 3.3.

4. Wire the circuit as shown in figure 3.9, but for now only connect the blue wire to *RS* and the white wire with the blue stripe to ground. For the moment leave the orange colored wire unconnected.

5. As a validation test, apply power and follow the procedures described in Experiment 3.2 to measure the plateau voltage shown on oscilloscope channel 2. Then calculate and record the impedance as you've done in previous experiments. This will be referred to as the *single line impedance*. The impedance should match the value you found when performing Experiment 3.2.

6. Now connect the orange colored wire to resistor *RS*2.If you're careful this can be done with the 5V power supply still applied. Observe that the plateau voltage displayed on oscilloscope channel 2 has increased. Record this value in your notebook and calculate and record the impedance. This is the *even mode impedance* because the two signal wires are switching in phase.

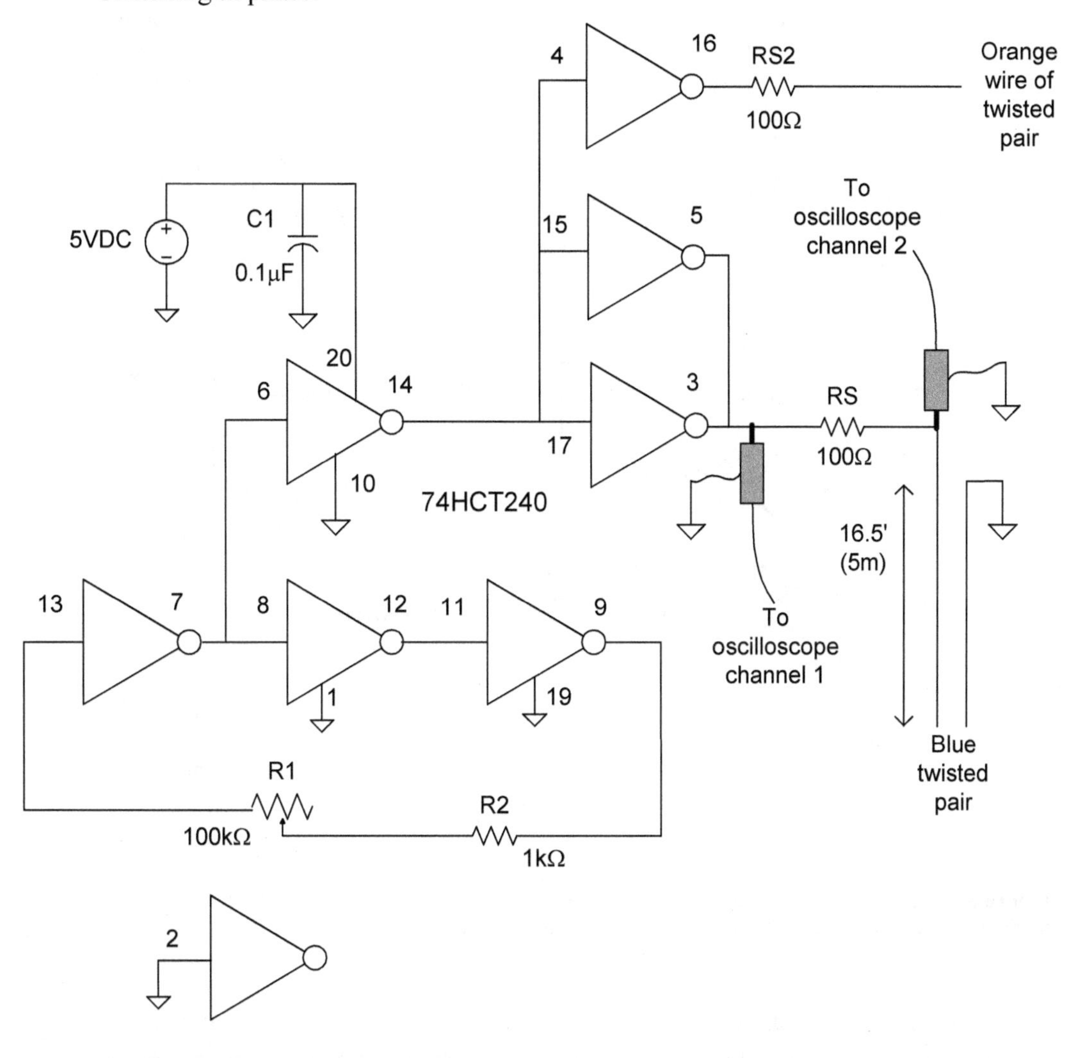

Figure 3.9 Test setup for the first part of Experiment 3.4 demonstrating in-phase ("even mode") switching (pin 16 has the same logical polarity as pins 5 and 3).

7. Switch off the power and rewire the circuit as shown in Figure 3.10. The added inverter causes the signal sent down the orange colored wire to be out of phase with the signal sent down the blue pair: Pin 16 will be driving a logic low when pins 3 and 5 are driving logic high levels, for example.

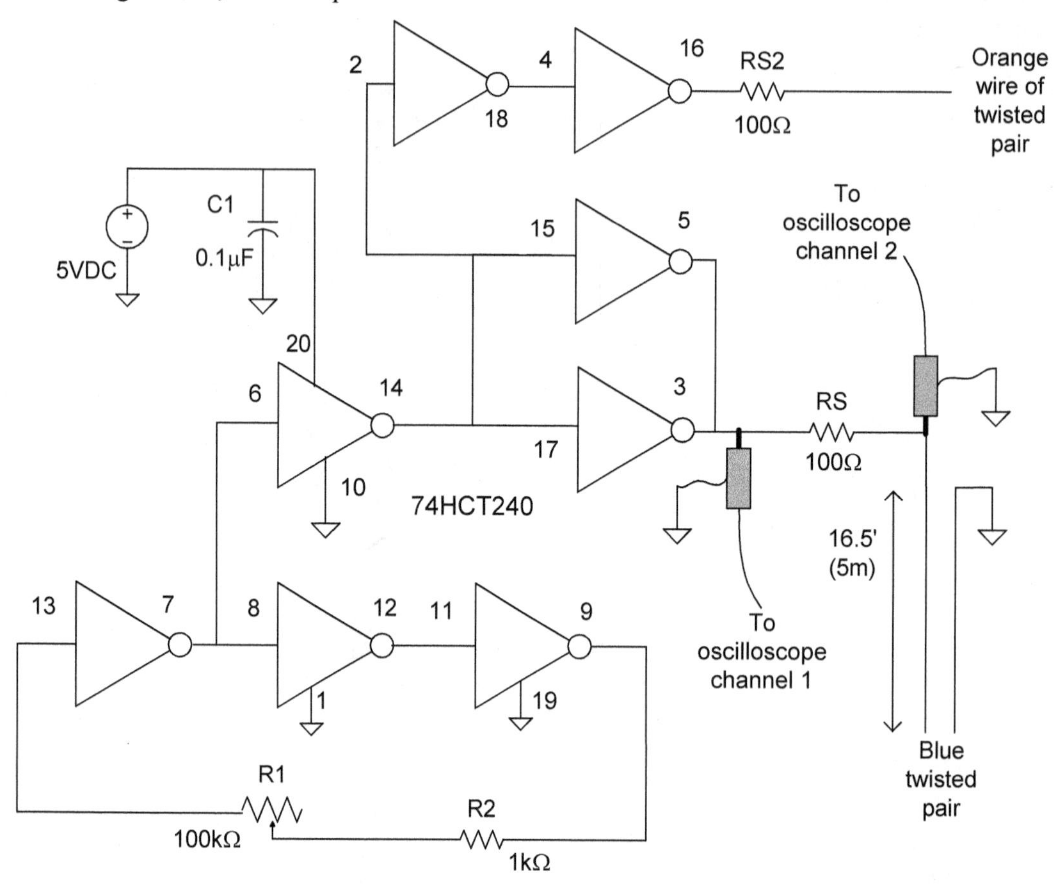

Figure 3.10 Test setup for the second part of Experiment 3.4 demonstrating out of-phase ("odd mode") switching (pin 16 has the opposite logical polarity as pins 5 and 3).

8. Apply power and measure the plateau voltage shown on oscilloscope channel 2. Notice it's lower than the previous voltages. Calculate and record the impedance in your notebook. This is the *odd mode impedance* because the two signal wires are switching out of phase.

Example Results

Figure 3.11 shows example waveforms and the results of the impedance calculations.

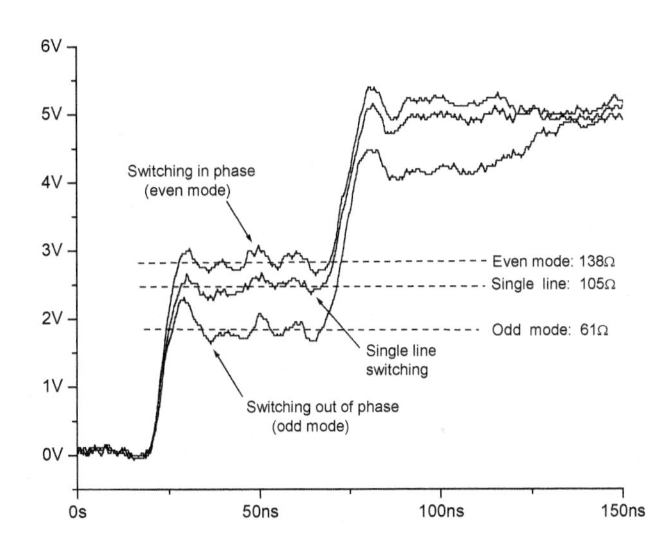

Figure 3.11 Test result showing odd and even mode impedances. Notice that the even mode impedance is higher than the odd mode impedance, and that the single line impedance value is between the two.

Comments

Although we often think the impedance of a signal wire as having a constant value, Figure 3.11 shows that if coupling is high enough the switching behavior of nearby wires will cause the impedance of the signal wire to change. The impedance is less in the odd mode case (*Zoo*), and is higher in the even mode case (*Zoe*). This phenomenon doesn't only occur with wires in a cable. Circuit board traces behave in this way too.

As you'll see in the next two chapters, to properly terminate a transmission line it's necessary to know its impedance. This has huge implications for system level designers because it correctly suggests that if coupling is high enough the transmission line can't truly be terminated with a single resistor[3].

The measured waveforms are noisy, making precise measurement of the impedance impossible. However, they are accurate enough to demonstrate even and odd mode behavior.

[3] Two coupled lines (including differential pairs) can be properly terminated with 3 resistors even when in those situations when where the change in impedance is large. See "Transmission Lines" in the bibliography for books discussing this. For instance, a detailed explanation is given in "Understanding Signal Integrity".

Supplemental Exercises

- Repeat this experiment but connect the white wire with the orange stripe to ground. Does this alter the odd or even mode impedances? Does adding additional ground wires help to control the impedance?
- Replace resistor *RS2* with a value similar to the value calculated for the odd mode impedance (61Ω in the example). How does this change the even, odd and single line waveforms? Why?
- Connect the orange wire directly to pin 16 (effectively setting *RS2* to zero ohms) and repeat the experiments. How has the even, odd, and single line waveforms changed? Why?
- Add a capacitor having a value from between 10 to 100pF from pin 18 to ground in Figure 3.10. First imagine the effects this would have on the waveform appearing on the orange colored wire, and then predict how this would affect the signal appearing on the blue colored wire.

Experiment 3.5: Measuring the Propagation Delay of a Transmission Line

Purpose

To measure the propagation delay (*td*) of transmission lines, and to introduce the concept of signal reflections.

Procedure

1. The test setup is shown in Figure 3.12. It's identical to the setup used in Experiment 3.2, except for the placement of the oscilloscope probes. If you have not already performed Experiment 3.2, then before proceeding with this experiment check the build notes appearing in Appendix A for construction details and circuit operation, and for information regarding parts substitution.

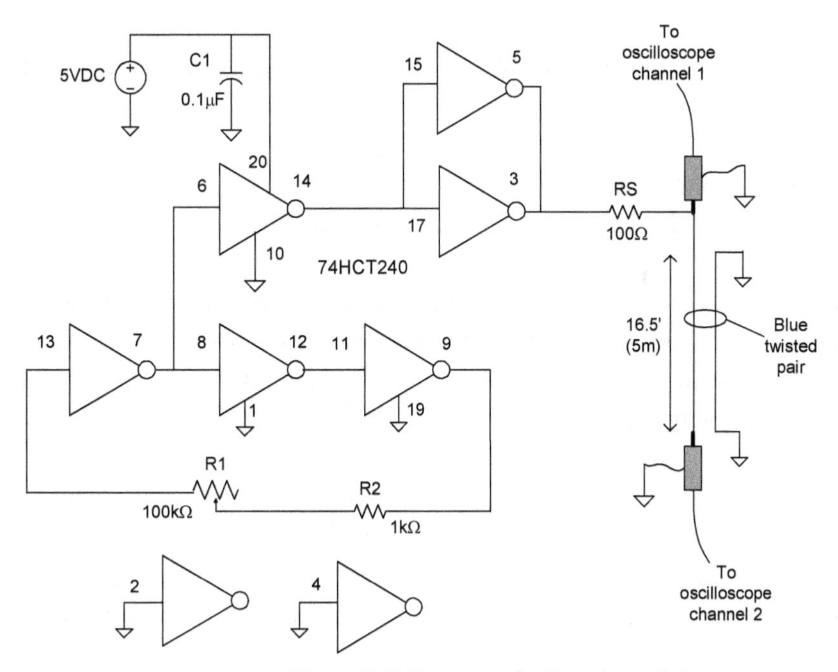

Figure 3.12 Test setup for Experiment 3.5.

2. Prepare 16.5 feet (5 meters) of cable as described in Appendix B. This has already been done if you've previously performed Experiment 3.2.
    a. In Experiment 3.2 you only connected the cable at the near end, but in this experiment you'll also be making connections at the far end. Untwist the blue pair of wires at the far end and remove about ¼ inch (0.6cm) of insulation from them.

3. Connect both ends of the white wire with the blue stripe to ground. Ideally the two ground connections should be close to one another (adjacent to each other on the breadboard), and close to pin 10 of the integrated circuit. Connect the ground clips for the two oscilloscope probes as close as possible to this same ground point.

4. Connect one end of the blue colored wire to resistor *RS* (this is the near end connection), and connect the oscilloscope probe going to channel 2 to the far end of the blue colored wire.

5. Set the oscilloscope to trigger on the rising edge of channel 1.

6. Apply 5V. If necessary, adjust resistor *R1* until the pulse on pin 3 of the integrated circuit is at least 200ns wide.

7. Observe the plateau voltage displayed on oscilloscope channel 1, and observe that the waveform displayed on channel 2 does not have a plateau, but is delayed in time.

8. Measure the delay from when signal appearing on channel 1 begins to rise to when the signal appearing on channel 2 begins to rise. This is *td*, the delay time of the transmission line. Record this value in your notebook. Figure 3.13 illustrates this for the example measurements made on the prototype.

9. Measure the width of the plateau by measuring the time between when the signal first begins to rise to the point after the plateau where it again starts to rise. The proper measurement points are illustrated in Figure 3.13. Record this delay time as *roundtrip* in your notebook. Compute the ratio of *roundtrip* to *td*, and record this value in your notebook.

10. Determine the delay per inch (or cm) by dividing the value you obtained for *td* by the cables length. Compare this with the value appearing in Chapter 2's Table 2.1.

Example Results

The following waveforms were obtained from the prototype when 16.5 feet (5m) of CAT5e cable was used.

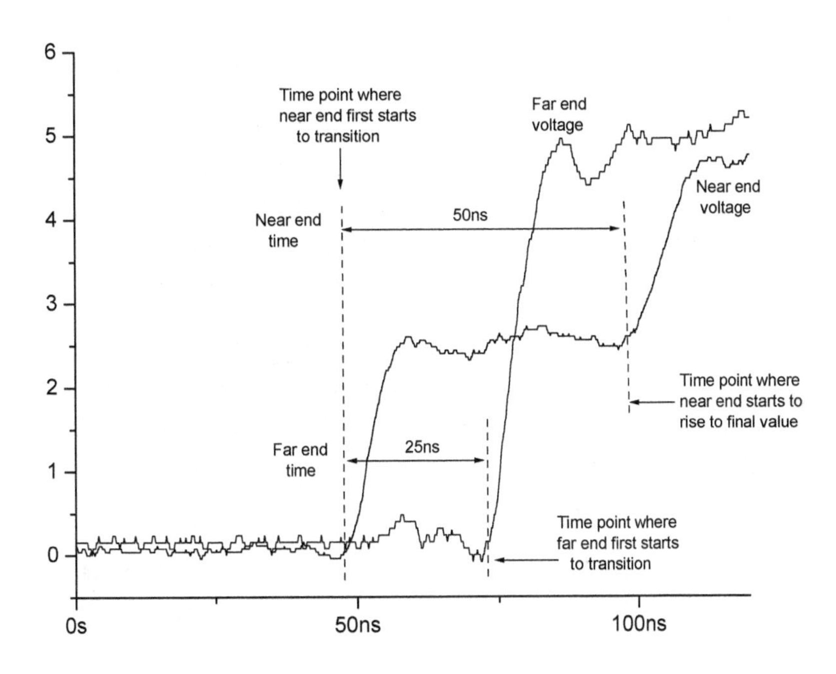

Figure 3.13 Experimental test results. Notice how signals are measured at their transition points.

Example Calculation

The time required for the signal to go from the near end to the far end is *td*: 25ns in the figure.

a) Since the cable is 16.5 feet (198 inches) long, the delay per inch is:

$$td/inch = \frac{25ns}{198inches} = 126ps/inch.$$

This compares very favorably with the specified value of 120ps/inch listed in Chapter 2's Table 2.1.

b) For the data in the example, the ratio between *roundtrip* and *td* is:

$$ratio = \frac{roundtrip}{td} = \frac{50ns}{25ns} = 2$$

Comments

It's not easy to precisely make these measurements, but the ratio you calculate should be close to 2, and the value of *td* you measure for your cable should be similar to the value in Chapter 2's Table 2.1.

Textbook timing is often specified as being measured between specific voltage levels where the logic changes state such as the 50% or 80% points. This strategy accounts for the signal rise time (the slope of the signal) and works well when it's necessary to determine the time delays in actual systems. We'll use this technique in Chapter 5 to measure the widths of reflected pulses.

The best way to get reliable and repeatable results when the signal transition has flat spots is to be consistent in where you measure the waveforms inflection points (the places where the waveform just starts to transition from one voltage level to another). These are marked in Figure 3.13, and it's evident from the figure just how subjective the process can be. However, you'll get reliable results so long as you're consistent in picking the same place on each waveform.

Notice how the width of the plateau is measured on channel 1. A common error is to measure just the flat portion and not to include the signal rise time. This mistake underreports the true width of the plateau by an amount roughly equal to the signal rise time. The correct method is to include the rise time by making the measurement at the inflection point, as is shown in the figure.

Supplemental Exercise

- If you are fortunate to have access to an RLC meter or bridge, use it to find the capacitance and inductance of the cable you used in this experiment. Then use equation (2.3) from Chapter 2 to calculate the delay. How does it compare to the values you measured experimentally?

Experiment 3.6: Finding the Capacitance and Inductance Once the Impedance and Delay are known

Purpose

To use measured values of the propagation delay and impedance of a transmission line to determine its capacitance and inductance.

Procedure

1. Determine the impedance and delay of a transmission line by using the techniques described in the previous two experiments.
2. Calculate the capacitance (*C*) and inductance (*L*) with the following formulas:

$$C = \frac{td}{Zo}$$

$$L = Zo \times td$$

These equations are derived from equations (2.2) and (2.3) in Chapter 2 by solving them simultaneously, first for *C* and then for *L*.

Example Calculation

From Figure 3.13, we know the one way delay of the line (*td*) is 25ns.

The measured impedance from Figure 3.12 is 105Ω.

The capacitance and inductance for the entire16.5 foot (5m) length of cable is:

$$C = \frac{td}{Zo} = \frac{25ns}{105\Omega} = 238pf$$

$$L = Zo \times td = 105\Omega \times 25ns = 2.6\mu H$$

These compare favorably to values measured with an RLC bridge of 250pF and 2.5μH.

Chapter 4 Overview of Reflections and Terminations

In this chapter you'll learn how signals sent down a transmission line travel as waves, and how this causes reflections to be created. You'll see how terminating the line by placing resistance at the drivers output (the near end), or at the receiver (the far end) can reduce or eliminate these reflections. Reflections can also be clamped (but not eliminated) by using diodes at the far end.

The Bibliography lists other books and papers that describe reflections and terminations in more detail than we do here.

Transmission Line Voltage and Current Waves

Chapter 2 showed that the amount of current initially sent down a transmission line depends on the applied voltage and the line's impedance. That makes sense from a circuit design point of view: When losses are low the line acts as a simple resistance and according to Ohms law, dividing the voltage by the resistance (which for a transmission line is the characteristic impedance) gives the current.

Although correct, this circuit approach prevents us from seeing a much deeper truth: The voltage and current sent down the transmission line travel as waves. The bibliography lists references that explore this in detail, but for the purposes of this manual we'll simply accept that the energy traveling along a transmission line does so as voltage and current waves.

The energy wave is called *electromagnetic* to signify its two parts: The voltage represents the electric part of the wave and the current the magnetic part. Capacitors model the way in which the transmission line reacts to the electric wave; inductors model the reaction to the current.

Like water waves, electromagnetic waves will be reflected when they encounter an appropriate obstacle. For water waves the obstacle might be a wall; for light waves a mirror. For the voltage and current waves travelling along a transmission line, the obstacle is a change in impedance.

We can see how a change in impedance might cause a reflection by considering what happens to the voltage wave when it reaches a short circuit at the end of a transmission line.

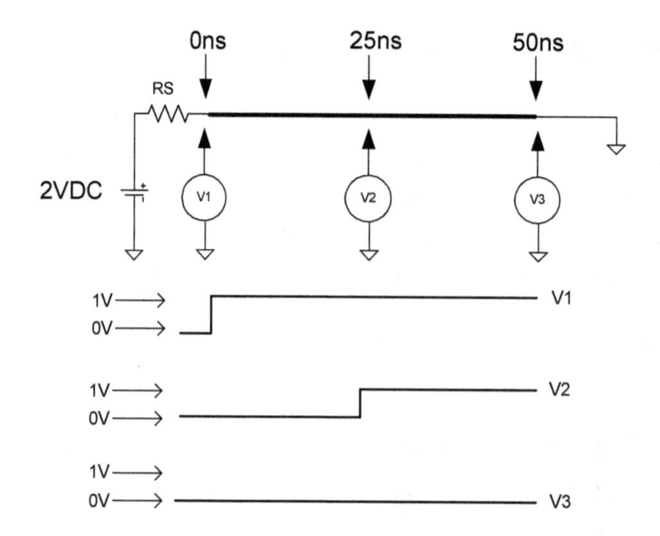

Figure 4.1 Connecting a battery to a transmission line causes a voltage wave to travel down its length. Voltmeters V1 and V2 show the wave's progress. The short circuit causes V3 to always read 0V.

In the figure a 2V battery has been connected through resistor *RS* to a transmission line having a delay of 50ns, which we recognize from Chapters 2 and 3 is the one way delay of the line, *td*. The far end is shorted to ground with a zero ohm jumper. Resistor *RS* has the same value as the transmission line impedance, so as Chapters 2 and 3 show, 1V will be launched down the transmission line from the 2V battery. The battery is putting put a constant 2V, not a pulse.

Voltmeters placed at the near end (the 0ns mark in the graph), the half-way point (the 25ns mark) and at the far end (the 50ns point) allow us to observe the voltage wave as it passes by.

This first test will run for 50ns. Once the results from that test have been examined we'll observe the behavior of the voltage wave over a period of 100ns.

For this first test we see from the figure that:

1. Voltmeter *V1* goes from 0V to 1V at time 0. It's still reading 1V at 25ns and 50ns. Evidently voltmeter *V1* doesn't 'know' about the short circuit present at the other end of the line, because if it did, *V1* would register zero at 50ns when the wave first reaches the end.
2. Voltmeter *V2* reads 0V until 25ns. This is the amount of time required for the voltage wave to travel from resistor *RS* to *V2*. The voltmeter then reads 1V, and that reading doesn't change when the wave reaches the end of the line at 50ns. Apparently *V2* is unable to sense the short circuit even when there has been enough time for the wave to reach the end of the line.

3. Voltmeter *V3* always reads 0V, even when the 1V wave reaches the end of the line at 50ns. This last observation makes sense because of the zero ohm connection to ground, but these results raise a question: Where did the 1V launched from the battery "go", and why didn't *V1* and *V2* read zero once the wave reached the short at the end of the line?

We can get more insight by extending the analysis out to 100ns (a round trip), as is done in Figure 4.2.

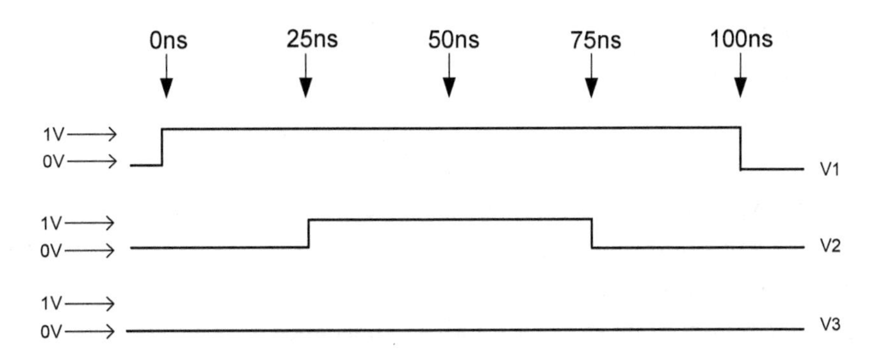

Figure 4.2 Behavior of pulses when examining the line for a round trip time.

Starting at 25ns, the voltage at *V2* remains at 1V until it goes to zero at 75ns, and the *V1* voltage remains at 1V until 100ns, where it too finally goes to zero.

To understand the behavior appearing in Figure 4.2 we first need to know how the electromagnetic wave behaves when encounters a change in impedance.

Rules for Reflections

The texts listed in the bibliography describe the wave differential equations verifying the following rules for reflections, but for the purposes of this manual we'll simply accept that waves are reflected as described below:

1. The voltage reflection will have a negative polarity if the resistance at end of the line has a value less than the impedance of the transmission line (such as a short circuit).
2. The voltage reflection will have a positive polarity if the resistance at end of the line has a value higher than the impedance of the transmission line (such as an open circuit).
3. The magnitude of the reflection depends on how much the impedance change differs from the transmission lines characteristic impedance. Larger changes cause larger reflections.

4. There are no reflections created when a resistor is placed at the end of the line having a value equal to the characteristic impedance of the transmission line. In effect, the line would look infinitely long in this case.
5. The reflection adds with the incident wave (actually, with any other waves present on the line). The voltage at any point along the line is the sum of all of the voltage waves present at that location.

Estimating the Reflection

The first four rules for reflections are captured below, in equation (4.1).

$$V_r = V_i \times \frac{RL - Zo}{RL + Zo} \tag{4.1}$$

Where:

- V_r is the value of the reflected voltage wave
- V_i is the value of the incident voltage wave
- RL is the resistance value of the load
- Zo is the transmission line characteristic impedance

The fifth rule is described by equation (4.2):

$$V_{total} = V_r + V_i \tag{4.2}$$

To get an idea of how to use these equations we'll test them against the above rules.

1. For a short circuit *RL* is equal to zero, which makes $V_r = V_i \times \frac{-Zo}{+Zo} = -V_i$: The reflected wave is equal to the incident wave, but (because of the minus sign) is of the opposite polarity. The voltage at the load is zero because: $V_{total} = V_r + V_i = -V_i + V_i = 0$
2. For an open circuit RL is infinite. Because *Zo* is so much smaller than infinity it can be ignored, and $V_r = V_i \times \frac{RL}{RL} = V_i$: The reflected wave is equal to the incident wave, and has the same polarity. The voltage at the load is twice the voltage launched down the line (it "doubles") : $V_{total} = V_r + V_i = V_i + V_i = 2V_i$
3. If a 250Ω resistor is placed at the end of a 50Ω transmission line the equation predicts a reflection of $V_r = V_i \times \frac{250 - 50}{250 + 50} = +0.67 \times V_i$. The voltage is 1 2/3's the value of the

launched voltage: $V_{total} = V_r + V_i = 0.67V_i + V_i = 1.67V_i$. This is 1/3ed lower than what the reflection would be if the line were an open circuit.

4. If a 50Ω resistor is placed at the end of a 50Ω transmission line the equation predicts a reflection of $V_r = V_i \times \frac{50-50}{50+50} = 0$: There is no reflection and the voltage at the load equals the launched voltage: $V_{total} = V_r + V_i = 0 + V_i = V_i$

Physical Explanation for the Example Reflection

Figure 4.3 is a version of the distributed circuit model introduced in Chapter 2. Because the assumption is that the line has no losses we're justified in simplifying the model by removing the resistors present in those models. We already know from the earlier chapters that an actual transmission line would consist of an infinite number of small capacitors and inductors, but for clarity the figure only shows them at 3 locations.

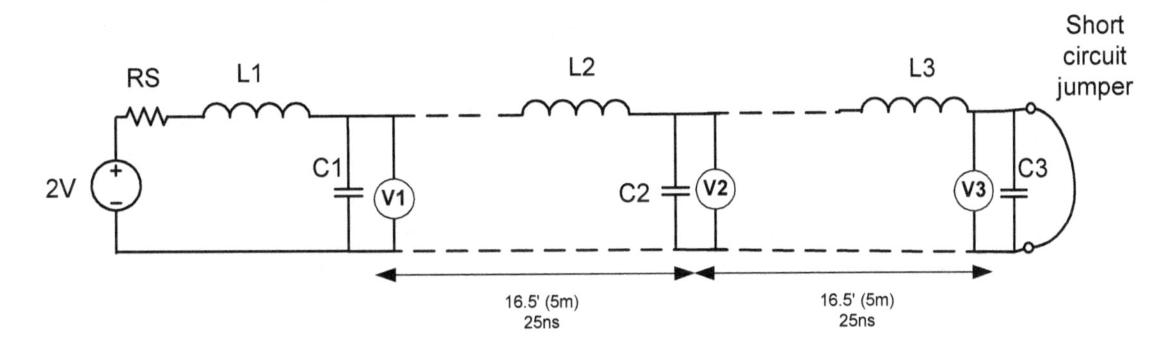

Figure 4.3 Transmission line model with no losses.

With Figure 4.3 and the rules for reflections in place we're now ready to explain the results from Figure 4.2.

1. Voltage divider action (which we discussed in Chapter 2, and experimented with in Chapter 3) causes the 2V battery to launch a 1V wave down the line (this is the *incident wave*). The incident wave charges *C1*to 1V, which is registered on voltmeter *V1*.
2. The 1V voltage incident wave travels down the transmission line, charging each incremental capacitor to 1V as it passes by.
3. After a delay of 25ns the incident wave has travelled exactly half way down the 50ns long line, and charges capacitor *C2* to 1V. Voltmeter V2 registers the wave's appearance.

4. The incident wave continues on, and 50ns after the voltage first appeared at *V1*, the incident wave's leading edge arrives at the end of the line where it encounters the short circuit. Since the 0 ohms of the short circuit doesn't match the impedance of the transmission line, a reflection is created.
    a. Voltmeters *V1* and *V2* still register the 1V present on charged capacitors *C1* and *C2*.
    b. In accordance with reflection rule number 1, for a short circuit the voltage reflection has the same magnitude (1V) as the incident wave, but of the opposite polarity. This makes the reflection -1V.
    c. The incident and reflected voltage waves combine and sum to zero, which sets the voltage cross capacitor *C3* to zero. Voltmeter *V3* registers this voltage.
5. The -1V reflection travels back up the line, toward resistor *RS*. As it does so it adds with the +1V that the incident wave had stored in each of the distributed capacitors, setting them to zero as it passes. The reflection reaches capacitor *C2* 25ns after it was created at the short circuit. To an outside observer this event occurs at an elapsed time of (50ns + 25ns) = 75ns (that is, the time required for one trip down the line plus the time required to travel half-way back up the line).
6. When the -1V reflection reaches *V1* it sets capacitor *C1*to zero, completing the cycle. The reflection required the same amount of time (50ns) to travel back up the line that the incident wave took to travel down it, so the reflection arrives at *C1* 50ns after it was created at the short circuit. The elapsed time is (50ns + 50ns) = 100ns (which you may recognize from Chapter 2 as being the round trip time). Even though the battery voltage remains constant at 2V, the voltage at node *V1* is a 1V pulse having a width equal to the roundtrip time of the transmission line. This matches the results shown in Figure 4.2.

Notice if *RS* had a value different from the impedance of the transmission lines characteristic impedance a new reflection would have been created at this point in the cycle. This new reflection would travel down the line toward *V2* and eventually be reflected by the short circuit at *V3* where the cycle would repeat. By making *RS* equal to the characteristic impedance of the line these types of multiple reflections don't occur, and is the principal governing *source series termination*.

Positive pulses are created at *V1* and *V2* by connecting the battery as shown in the figure and keeping the 2V applied long enough for the reflection to make it back to *RS* (a round trip time). The pulses would be negative (having values lower than zero volts) if the battery voltage was switched from 2V back to zero volts. Next chapter's experiments explore the results when the connection at *V3* is an open rather than a short circuit, when resistor *RS* has a value that's different from the transmission lines characteristic impedance, and when a resistor is connected at *V3*.

Reflections in Practical Systems

Resistors, logic gates and connections in practical systems aren't perfect: They have some inductance and capacitance. This is one reason the pulses you measure in actual hardware won't look as clean and unambiguous as those shown in Figures 4.1 and 4.2. The rise time of the incident wave further complicates the waveform, as you'll experience firsthand when performing the next chapter's experiments.

Capacitors and Inductors as Loads

As Figure 4.4 shows, imagine that the short circuit at *V3* is replaced with a capacitor load, which we'll call *CL*. The inputs of many CMOS logic devices can be modeled as a lumped capacitor, especially at low frequencies. Load capacitor *CL* is initially discharged to zero volts, so a pulse encountering the capacitor would at first see a short circuit and then eventually an open circuit once the capacitor had fully charged. This is evident in the plot appearing in Figure 4.5 where *V1* is the launched voltage and *V3* the voltage at the load.

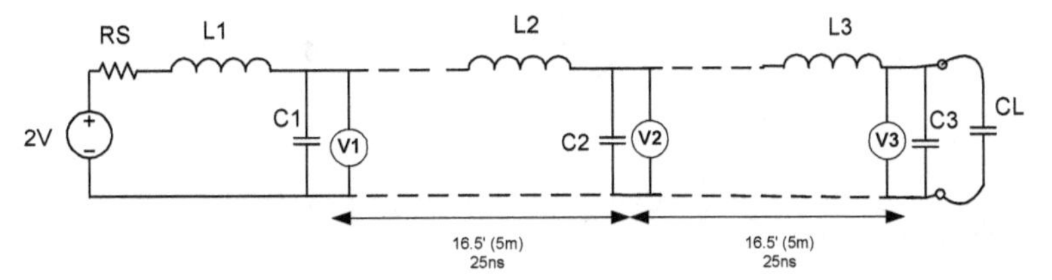

Figure 4.4 Lossless transmission line with a capacitor load.

From the figure we see how the voltage at *V3* is zero for the first 50ns, which is the time the signal requires to travel the length of the line and appear a capacitor *CL*. It then rises from zero volts to 2V with the classic exponential curve typical of series RC charging circuits. The important observation here is that the reflection generated by *CL* takes on this same shape.

The voltage across *CL* eventually reaches 2V because for an open circuit (which is what the capacitor becomes once it has been fully charged by the wave) the reflection has the same magnitude and value as the incident wave. Signal integrity engineers often refer to this as *doubling*, because the voltage becomes twice the value of the launched voltage (which was 1V in this case).

The first portion of the reflection arrives at V1 at precisely 100ns. Since *CL* was initially a short circuit to ground, *V1* is first driven to nearly zero volts (it would have been precisely zero volts if the capacitor was a perfect short circuit) and then rises exponentially toward a final value of 2V.

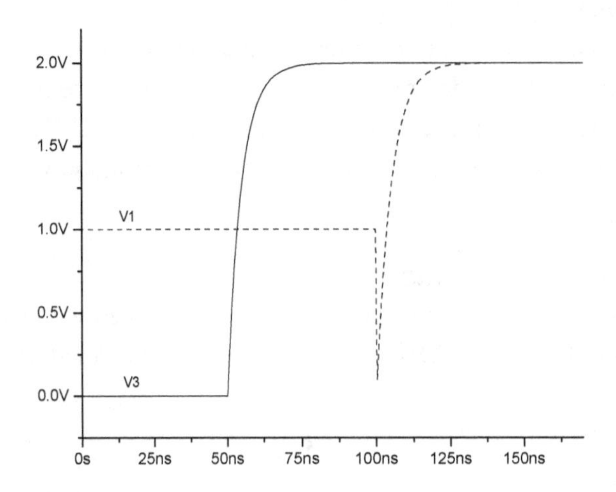

Figure 4.5 Response with a capacitor load. The voltage at the launched end is V1; V3 is the load end signal (corresponding to Figure 4.4).

Although not shown, an inductor works in the opposite way: The incident wave would initially see an open circuit, which then becomes a short over time. In these circuits the voltage at *V3* starts at zero volts, doubles (in the ideal case the inductor initially is a perfect open circuit) and then exponentially decays back to zero. For either capacitor or inductors, the *V1* voltage has the same shape as the reflection from *V3*.

These observations have important ramifications for properly terminating transmission lines, as is discussed next.

Terminations

Terminations are used to prevent reflections, or to reduce their effects. In the next chapter you'll experiment with three of the most common termination schemes used in systems employing CMOS logic: Source Series, Far End Resistance, and Far End Diode Clamps.

Three other termination types (Voltage divider, Thevenin and AC terminations) are not presented in this manual, but they are described in the books listed in the bibliography.

Source Series Termination

In Source Series termination a resistor (*RS*) is placed in series with the output of the logic gate. Adding a resistor in series with the driver as is shown in Figure 4.6 changes its operation in two ways.

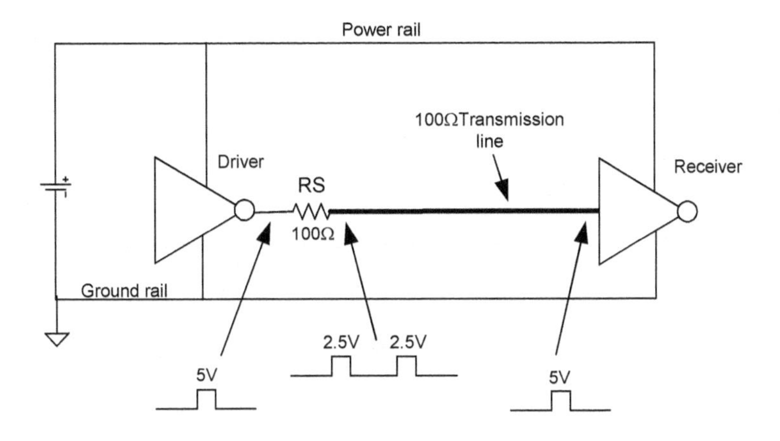

Figure 4.6 Adding resistor RS at the output of the driver source series terminates the transmission line.

First, the added resistance effectively increases the drivers "on" resistance (RDS_on). This lowers the current the driver can send down the transmission line, which is the same thing as reducing the driver's strength. This strategy is very helpful when the driver's "on" resistance is much lower than the transmission lines characteristic impedance (*Zo*). Adding just enough resistance so that the sum of the "on" resistance of the driver and the value of *RS* equals *Zo* causes the driver to be matched to the line, but this doesn't stop a reflection from being created at the load. In fact, source series termination relies on the load reflection to sum with the incident wave, creating a pulse having the proper voltage. For instance, as explained and demonstrated in Chapters 2 and 3, a 5V driver in series with *RS* of 100Ω launches 2.5V down a 100Ω transmission line. This incident voltage wave adds to the 2.5V reflection wave it creates at the receiver, forming a 5V pulse.

The 2.5V reflection then travels back up the line toward *RS*, which brings us to the second important operational characteristic of source series termination: The 2.5V reflection can be seen when it arrives back at *RS* (this accounts for the two 2.5V pulses shown in the figure), but it won't re-reflect if the sum of the "on" resistance and *RS* is equal to *Zo*. You'll see in the next chapters experiments how, if left unchecked, re-reflections can cause very distorted looking signals at the load. Adding a series resistor close to the value of *Zo* will greatly reduce (and can fully eliminate) the creation of these re-reflections. This can improve signal quality significantly, but you'll also see from the experiments how the signal can be distorted if *RS* is made too large.

Far End Resistance Termination

Where source series termination uses the reflection at the load to increase the voltage at the receiver, far end termination (shown below, in Figure 4.7) prevents that reflection from occurring in the first place.

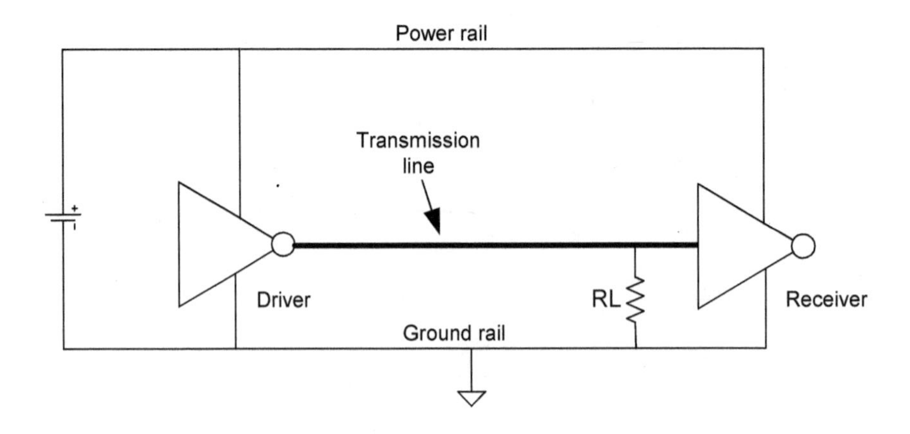

Figure 4.7 Adding resistor RL at the end of the transmission line, near the load, creates parallel termination.

For this scheme to work the output resistance of the driver (RDS_on) must be low enough to launch a voltage that's sufficiently high to be recognized by the receiver as a valid logic level without requiring a boost from a reflection.

Because the driver impedance is so low, any reflection generated at the load that finds its way back to the driver will be re-reflected, distorting the signal. The reflection isn't created (and so, can't be re-reflected) when the impedance of the load equals the impedance of the transmission line. For this reason, the value of the termination resistor *RL* is usually made equal to the value of the transmission line impedance, but as experiment 5.6 shows, it can be advantageous to use a different value.

Resistor *RL* will work just as well if it's connected to the power rail rather than to ground, provided adequate decoupling capacitance is connected between the rail and ground. For this to be most effective the capacitor must be placed physically close to *RL*. Connecting *RL* to power provides a "pull-up" configuration where the gate input is pulled up if the transmission line is disconnected (or if the driver gets turned off or becomes disconnected).

It's also possible to terminate the line by simultaneously using a pull-up and a pull-down resistor (a pull-down is a resistor to ground, as is shown in Figure 4.7). The resistors are chosen in value such that the resistance of the parallel combination equals the transmission line impedance, while the combination simultaneously creates a specific termination voltage. For instance, an 180Ω pull-up and a 220Ω pull-down has an equivalent termination resistance of very nearly 100Ω (which is good for termination many multi-wire cables) and when connected to a 5V power supply creates a 2.75V pull-up voltage. The bibliography lists references that describe this scheme in more detail.

Far end termination is very effective in reducing reflections, and so greatly reduces overshooting. However, it has one drawback: The terminating resistor is also a load that draws current from the driver. This is very different from source series termination where the driver only sources current until the reflections have died out (usually one round trip time). For this reason far end termination isn't usually used in low power designs.

Far End Diode Clamps

Diodes are sometimes placed at the transmission lines far end, at the load (at the receiver), to reduce or eliminate overshooting. The circuit is shown in Figure 4.8, where we see that one diode is connected to the power supply voltage, while a second is connected to ground. This creates a clipping circuit that prevents the signal from going more positive than the power supply voltage plus a diode drop, or more negative than a diode drop below ground.

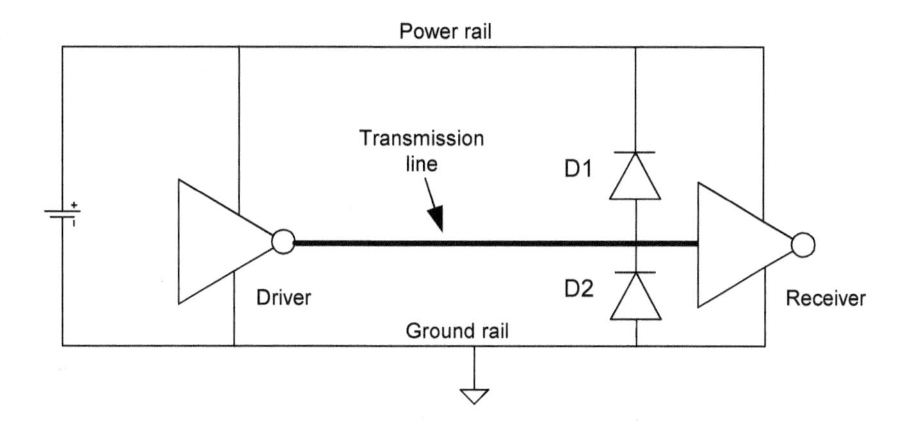

Figure 4.8 Clamping reflections by placing diodes at the transmission lines end.

In the experiments the power supply is set to +5V, and D1 and D2 are small signal silicon diodes having a voltage drop of 0.6V. This sets the theoretical clipping range for the signal to be between +5.6V and -0.6V. As you'll see in the experiments, in practice series resistance and inductance allow the signal to exceed these bounds.

Provided they are connected to the same power and ground as the driver, diode clamps have an advantage over far end resistor termination in that current only flows when they are clamping. When they do turn on they redirect overshoot energy into either the power supply or ground system. Managing this energy can require careful attention, especially in small, low power systems. Adequate decoupling capacitance must be in place near D1 and D2 for them to effectively clamp the overshoots without causing power supply noise, and the power and ground connections must have low inductance. This noise can show up as glitches or spikes on the power wiring, and it can lead to RF noise (EMC compliance) problems. Low inductance decoupling capacitance near D1 and D2 can help reduce this noise.

Chapter 5 Laboratory Exercises: Reflections and Terminations

Reflections and the way in which different types of terminations change reflections are observed in this chapter. For best understanding you should perform the Chapter 3 experiments first, before you attempt these experiments.

Experiment 5.1: Effects of Capacitor Load on Reflected Waveform

Purpose

To observe the reflections created by capacitors. This is important because the inputs of most CMOS logic devices act as small capacitors, especially at low frequencies.

- Note: The test setup described here is identical to the setup used in Experiment 3.5, except for the addition of load capacitor *CL*. Follow the Experiment 3.5 build instructions and perform that experiment before proceeding to this one. Appendix A describes suitable capacitor types for *CL*.

Procedure

1. Follow steps 1 – 5 of Experiment 3.5.
2. Add capacitor *CL* to the Experiment 3.5 test setup, as shown in Figure 5.1.

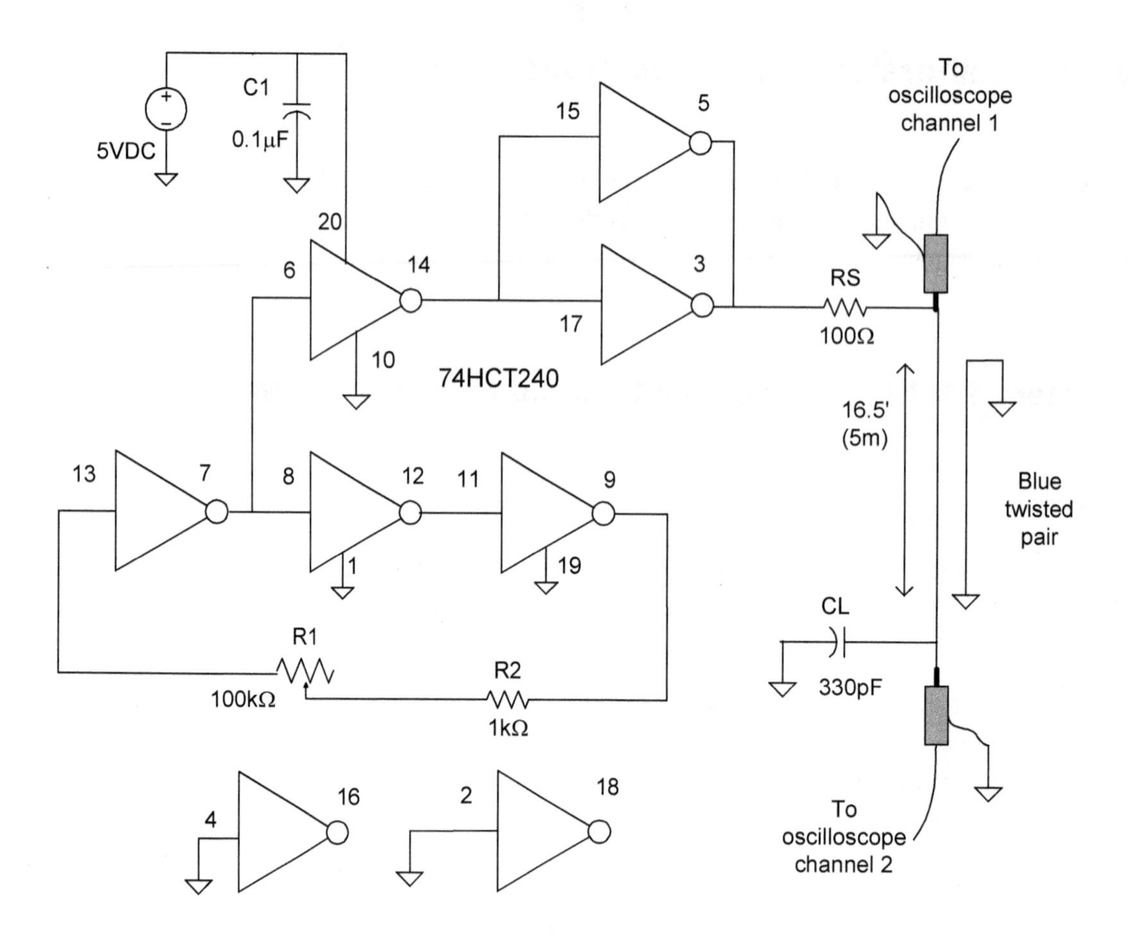

Figure 5.1 Test setup for Experiment 5.1. This is identical to Figure 3.12 but with the addition of capacitor CL.

3. Apply 5V to the circuit. If necessary, adjust resistor *R1* so that the pulse appearing on pins 3 and 5 of the integrated circuit is 200ns or wider. The actual value isn't critical.
4. Observe the waveforms at the near end (oscilloscope channel 1) and far end (channel 2). Set the oscilloscope time base so that the rising edge displayed on channel 2 occupies a large portion of the screen. Figure 5.2 shows a sample waveform from the prototype hardware.

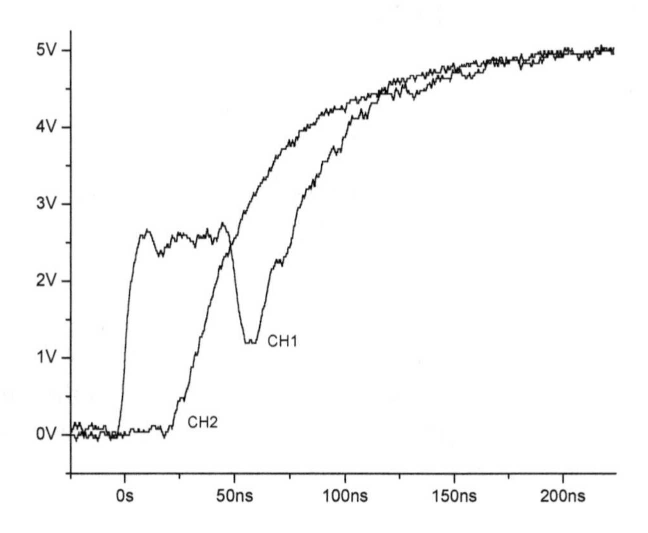

Figure 5.2 Capacitor CL causes a glitch and then an exponential rise in the channel 1 (the near end) waveform. Channel 2 shows the far end waveform.

Comments

Adding capacitor *CL* to the end of the line has changed the signal rise time and has added a negative going glitch to the signal observed at the near end. The channel 1 measurements (the near end measurements, that is, the launched signal) are very revealing: The initial rise time of the signal is very sharp, but it becomes much slower at about 60ns. The initial sharp rise occurs because the waveform hasn't yet experienced load capacitor *CL* (more to the point, the reflection from *CL* hasn't yet made its way back to the near end). The exponential charging of *CL* causes the slower rise time after 60ns has elapsed. The waveform initially spikes down (occurring at about time = 50ns) because a discharged capacitor initially "looks like a short circuit". We know that if *CL* was replaced with a short circuit the near end voltage would go to zero after a roundtrip time (50ns).

Supplemental Exercises

- Try different values for *CL*, and see how they alter the channel 1 and channel 2 waveforms.
- The input capacitance of many types of logic gates generally ranges from 5pF to 15pF. How does the waveform appear if you make *CL* 10pF?
- How does the waveform appear when a 100Ω resistor is placed in series with *CL*?
- How would the waveform appear if *CL* was replaced with an inductor?

Experiment 5.2: Far-End Resistor Termination

Purpose

To observe the effects of reflections when the driving impedance is low and the line is unterminated, and then observe the improvement when the line is terminated with various values of resistances.

- Note: The test setup described here is similar to the setup used in Experiment 3.5, except in this experiment resistor RS is deleted and load resistor *RL* is added. Follow the build instructions described for Experiment 3.5. Appendix A describes suitable resistor types for *RL*.

Procedure

1. The test setup is shown in Figure 5.3. Notice that unlike in previous experiments here the transmission line is directly connected to pins 3 and 5 of the inverter. Be sure to check the build notes in Appendix A for construction details and circuit operation, and for information regarding parts substitution. If necessary, review Appendix C for ways to use your oscilloscope to obtain clean waveforms.

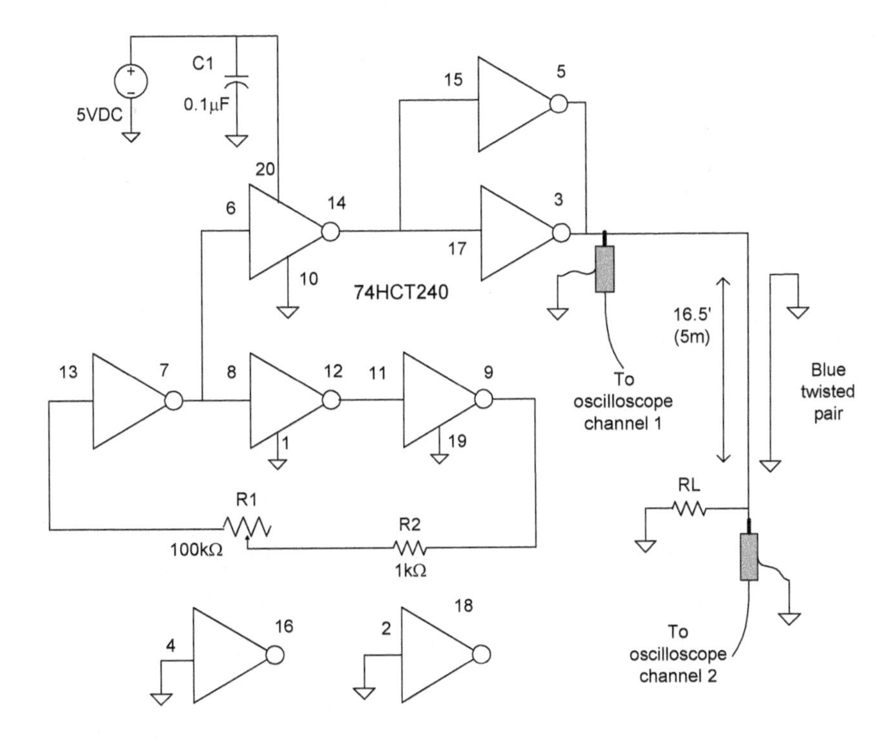

Figure 5.3 Hardware test setup for Experiment 5.2.

2. Do not yet connect resistor *RL*. It will be connected later, in step 6.
3. Connect the oscilloscope probes as shown in the figure.
    a. Set the trigger to occur on the rising edge of channel 1.
    b. You'll get cleaner, more easily measured waveforms when the oscilloscope probe's ground connection is made directly to pin 10 of the integrated circuit, and the probe ground lead is kept short. For lowest noise connect both of the twisted pair grounds as close to pin 10 as practical.
4. Apply power to the circuit and adjust resistor R1 so that the pulse appearing on channel 1 is at least 200ns wide. The actual width isn't critical.
5. Observe the voltages on oscilloscope channels 1 and 2. Figure 5.4 shows an example result.

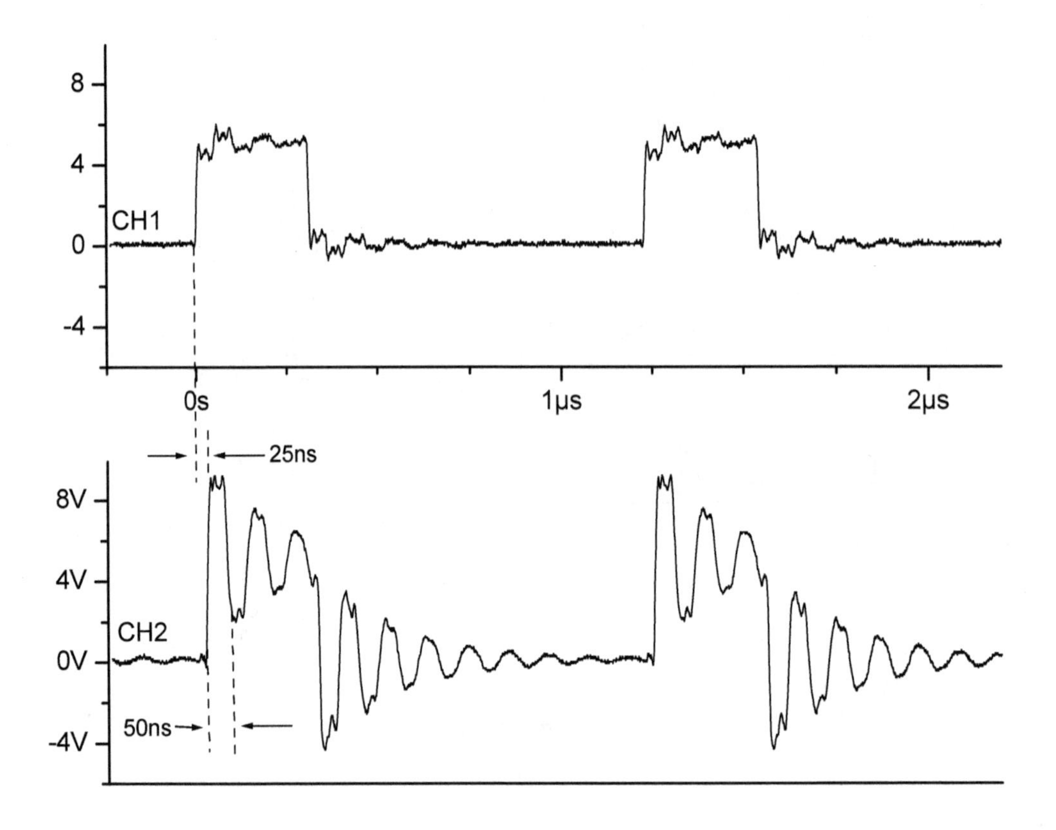

Figure 5.4 Oscilloscope waveforms from Experiment 5.2, step 5 (no termination case).

6. Use a 100Ω resistor for RL, and connect it as shown in Figure 5.3. Observe the voltages on channels 1 and 2. Figure 5.5 shows an example result. To make comparison easier, this data is displayed with the same scale as Figure 5.4.

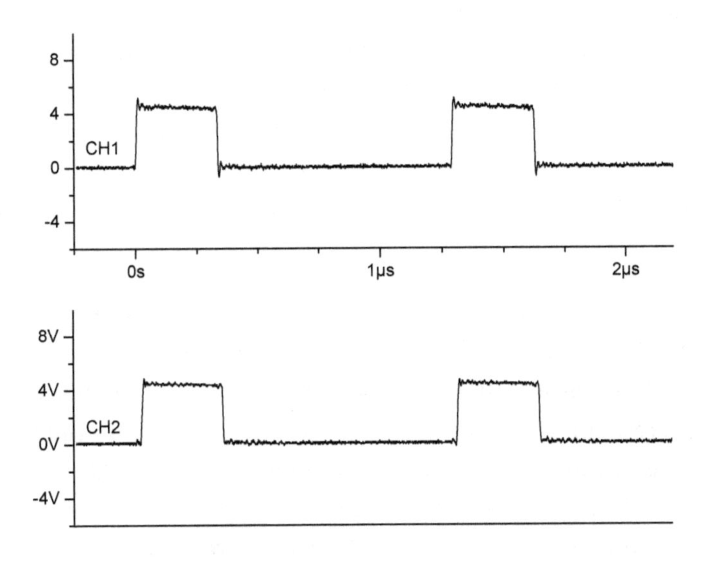

Figure 5.5 Oscilloscope waveforms from Experiment 5.2, step 6 (resistor termination at the far end).

Comments

Notice that the voltage launched from pins 3 and 5 (the incident wave, shown in Figures 5.4 and 5.5 on oscilloscope channel 1) is close to 5V in the figure, which is nearly twice as high as what was measured in Chapter 3's experiments. This comes about because those experiments included series resistor *RS*, which adds to the resistance of the integrated circuit, raising the drive impedance. Removing resistor *RS* and driving the transmission line directly from the integrated circuit as you are doing in this experiment results in the drive impedance being much lower than in the earlier experiments. This lower impedance is why the launched voltage and current waves have larger amplitudes here as compared to earlier experiments. Because the incident wave is so big in this example we should expect any reflections that get created to be larger than in previous experiments. More importantly, any reflections that do make it back to pins 3 and 5 will be re-reflected rather than being absorbed as they would have been had resistor *RS* been present, and this is quite clear from the waveform displayed in Figure 5.4 on oscilloscope channel 2. The peaks are high in amplitude and have widths of 50ns (which is the round trip time of the cable being used in these experiments). Notice too how losses cause the amplitude of each overshoot to be lower than the one occurring before it (that is, the ringing height is less over time).

The reflections are negative (have values less than 0 volts) when the signal switches from +5V to 0V (this occurs at about 300ns and 1.5µs in the figure), but otherwise the ringing from the falling edge has the same general characteristic as the rising edge ringing.

Adding the resistor to ground in step 6 terminates the line which results in the much cleaner waveforms at both the near and far ends which are shown in Figure 5.5. Notice that this improvement occurs even though the impedance at the near end of the line is not the same as the impedance of the transmission line. This must mean that the 100Ω resistor has prevented a reflection from occurring, otherwise ringing would have been present in the channel 2 waveform as the refection bounced off the low impedance present at pins 3 and 5 (as is shown in Figure 5.4).

Critical Observations

⇒ **Without proper termination, reflections can cause voltage pulses at the far end to be greater than the power supply voltage, and lower (more negative) than ground. The reflection pulse width is the same as the round trip delay of the line.**

⇒ **Placing a resistor at the far end that has a value equal to the transmission line characteristic impedance terminates the line and prevents reflections from occurring. "Matched far end termination" (which includes parallel termination) is the only termination technique that prevents reflections.**

Supplemental Exercises

- Repeat step 6 using different values for *RL*, ranging from about 47Ω to 1kΩ. Observe the effects these various resistances have on the waveforms. How important is it that *RL* exactly matches the transmission line impedance?
- Set *RL* to zero ohms (a short circuit to ground) and repeat step 6. How does this change the waveform? Compare these results to the discussion in Chapter 4.
- Repeat this experiment but disconnect *RL*'s ground connection and attach it to +5V instead. How does this alter the signal?
- Increase the resistance of *RL* to 220Ω, and add a second 220Ω resistor to the +5V supply. This forms a parallel terminator having an impedance equal to the parallel value of the two resistors -- 110Ω in this case. How has this changed the shape of the signals on oscilloscope channels 1 and 2?

Experiment 5.3: Far End Diode Clamp

Purpose

To observe how two diodes can be used to limit the magnitude of reflections.

- Note: Be sure to perform Experiment 5.2 at least up to step 5 before performing this experiment. This will allow you to see the differences between an open line and the effects when the line is clamped with diodes.

Procedure

1. The test setup is shown below, in Figure 5.6. This setup is identical to the setup used in Experiment 5.2, except resistor *RL* is replaced with two diodes. Type 1N4148 small signal diodes were used in the prototype, but other small signal diodes such as a 1N941, 1N457 or the 1N4454 are acceptable. The Schottky BAT41 is also a good choice. Review the build notes in Appendix A for construction details, and if necessary, review Appendix C for ways to use your oscilloscope to obtain clean waveforms.

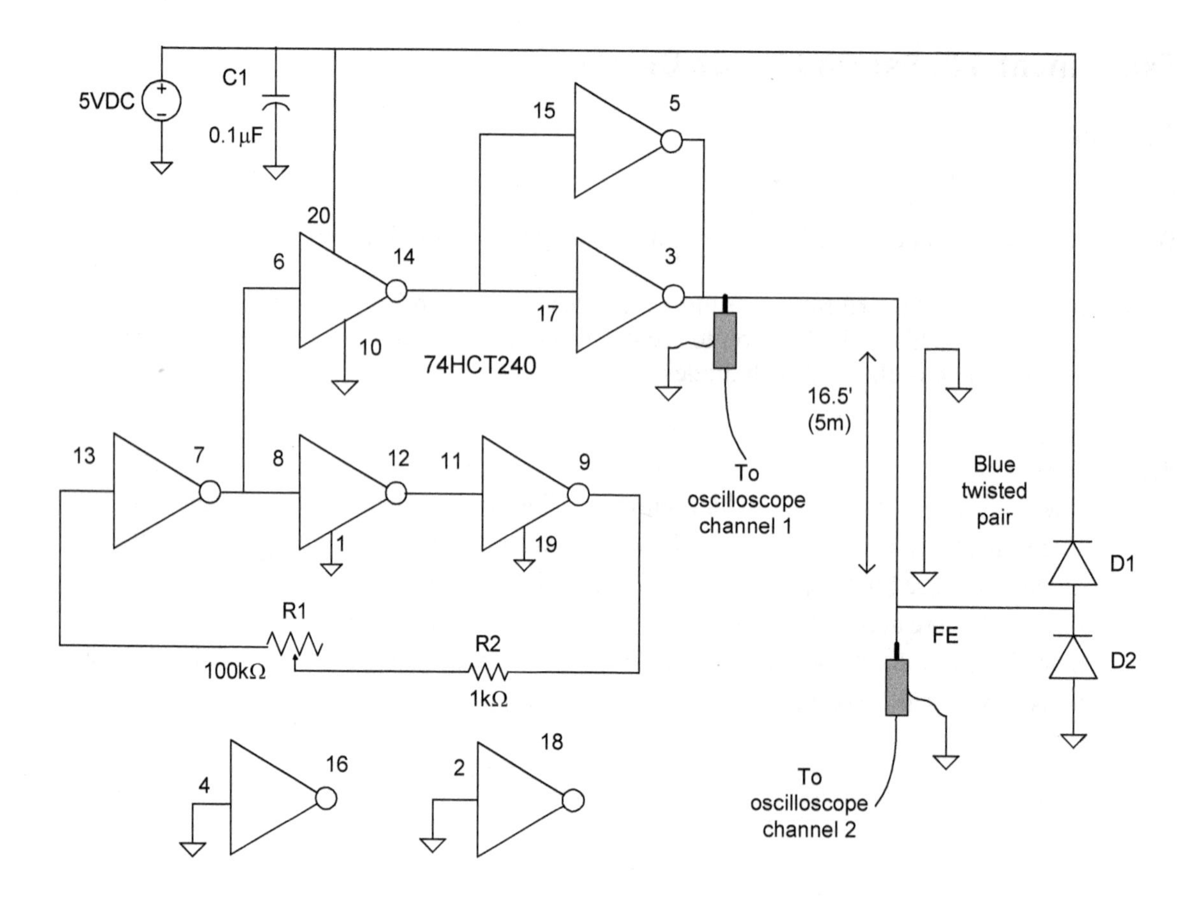

Figure 5.6 Test setup for Experiment 5.3.

2. Connect the oscilloscope probes as shown in the figure.
   a. Set the trigger to occur on the rising edge of channel 1.
   b. You'll get cleaner, more easily measured waveforms when the oscilloscope probe's ground connection is made directly to pin 10 of the integrated circuit, and the probe ground lead is kept short. For lowest noise connect both of the twisted pair grounds as close to pin 10 as practical.
   c. You'll get the best results by keeping the leads of *D1* and *D2* short.
3. Apply power to the circuit and adjust resistor *R1* so that the pulse appearing on channel 1 is at least 200ns wide. The actual width isn't critical.
4. Observe the voltages on channels 1 and 2. Figure 5.7 shows an example result.

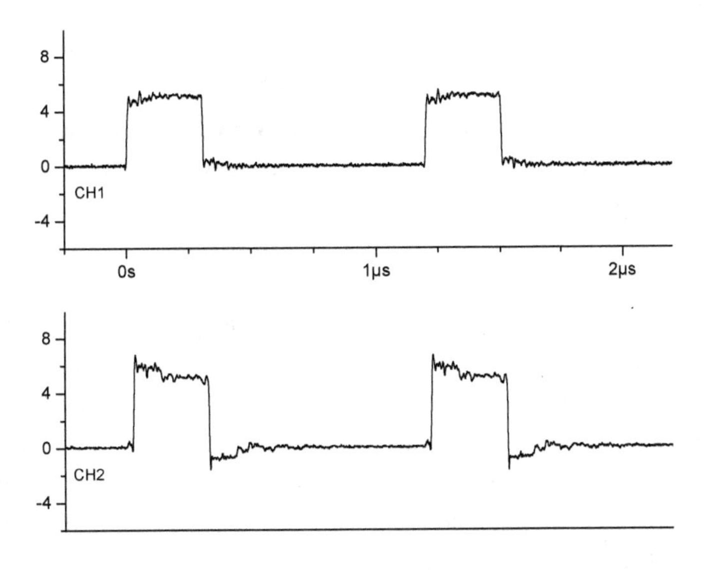

Figure 5.7 Oscilloscope waveforms from Experiment 5.3. Channel 1 shows the near end waveform. Channel 2 shows the waveform at the far end (*Vfe*).

Comments

As is evident from the CH2 waveform in the figure, placing diodes *D1*and *D2* at the end of an open line (or a line where the load is a high impedance) can be very effective in limiting ringing.

This technique is sometimes called "diode termination", but this term isn't correct. To "terminate" the line at the far end means an impedance that matches the transmission line impedance is placed at the load, preventing a reflection from being created. Diodes *D1* and *D2* work differently: They don't prevent reflections from being created. They simply prevent them from becoming too large. This is called "clipping", or "clamping", and the *D1/D2* circuit is properly called a "diode clamp".

Diode *D1* turns on when the voltage at node FE is higher than the voltage of the 5V supply, plus a diode drop (typically 0.6V). Diode *D2* turns on when the voltage is more negative than ground, minus a diode drop. This means that the voltage at node FE (which is oscilloscope channel 2) should never exceed +5.6V or -0.6V, but parasitic resistance of the diodes, and inductance in the wiring cause the clamping action to occur at voltages outside this range. This is evident in the bottom portion of Figure 5.7, where the voltage peaks extend to just over +6V and to -2V.

Critical Observations

- ⇒ **Diode clamps don't prevent reflections from occurring, but they do place limits on how high the reflection voltage can reach.**
- ⇒ **Unintended wiring inductance and the intrinsic on resistance of the diode will make the voltage at which clipping occurs greater than a nominal diode drop above the power supply rail.**

Supplemental Exercises

- Try different diodes for *D1* and *D2* (for instance, a 1N4001 silicon rectifier diode, or a BAT41 Schottky barrier diode). How are the waveforms changed? Why?
- Add between 1 and 30Ω of resistance in series with *D1*, and with *D2* to show how a diode's parasitic series resistance affects clamping. How do the waveforms change?
- Connect diode *D1* to a separate power supply, and adjust it to verify that a voltage lower than +5V results in less high-side ringing. What happens when the supply is 4.4V or lower? Why isn't using another power supply for only this purpose, (or a resistor voltage divider to create such a supply) common practice?
- Analyze where the current flows in the clamp circuit. How does this differ from a single terminating resistor to ground? Consider the magnitude and the direction of the current flow. What does this tell you about the proper ratings that a clamp diode should have?
- Analyze what would happen if *D1* and *D2* were placed at the near end (at pins 3 and 5 of the integrated circuit) rather than at the far end.

Experiment 5.4: The Effect of Far End Termination on a Gate Load

Purpose

To observe how the input of a CMOS gate clamps the signal waveform, with and without proper resistor termination.

- Note: Be sure to perform Experiment 5.2 at least to step number 5 before performing this experiment. This will allow you to see the differences between an open line and the effects when the line is connected to a gate load.
- It's helpful to have performed Experiment 5.3 before you begin this one so you can compare the effects of the parasitic diodes naturally present on the input gates of integrated circuits and diode clamps externally added to the circuit.

Procedure

1. The test setup is shown in Figure 5.8. This setup is identical to the setup used in Experiment 5.2, except in this experiment pin 2 of the integrated circuit is connected to the far end of the transmission line. Be sure to check the build notes in Appendix A for construction details and circuit operation, and for information regarding parts substitution. If necessary, review Appendix C for ways to use your oscilloscope to obtain clean waveforms.

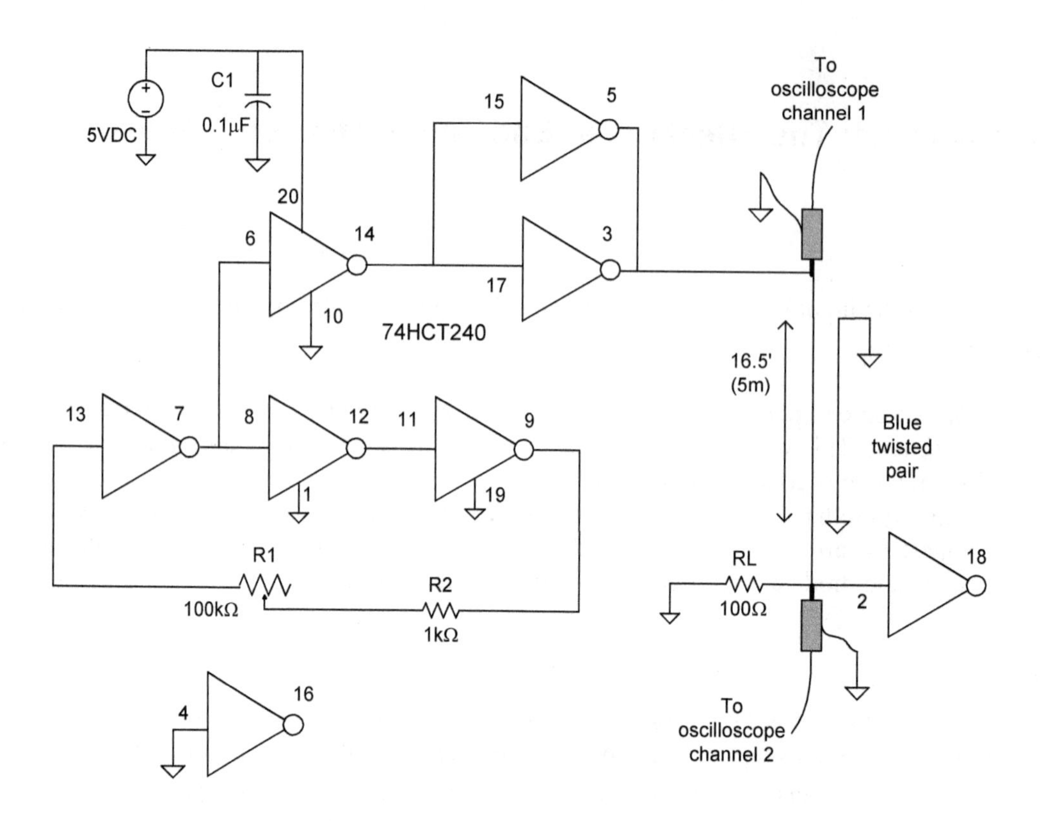

Figure 5.8 Test setup for Experiment 5.4.

2. Do not yet connect 100Ω load resistor *RL*. It will be connected later, in step 6.
3. Connect the oscilloscope probes as shown in the figure.
    a. Set the trigger to occur on the rising edge of channel 1.
    b. You'll get cleaner, more easily measured waveforms when the oscilloscope probe's ground connection is made directly to pin 10 of the integrated circuit, and the probe ground lead is kept short.
4. Apply power to the circuit and adjust resistor *R1* so that the pulse appearing on channel 1 is at least 200ns wide. The actual width isn't critical.
5. Observe the voltages on oscilloscope channels 1 and 2. Figure 5.9 shows an example result.

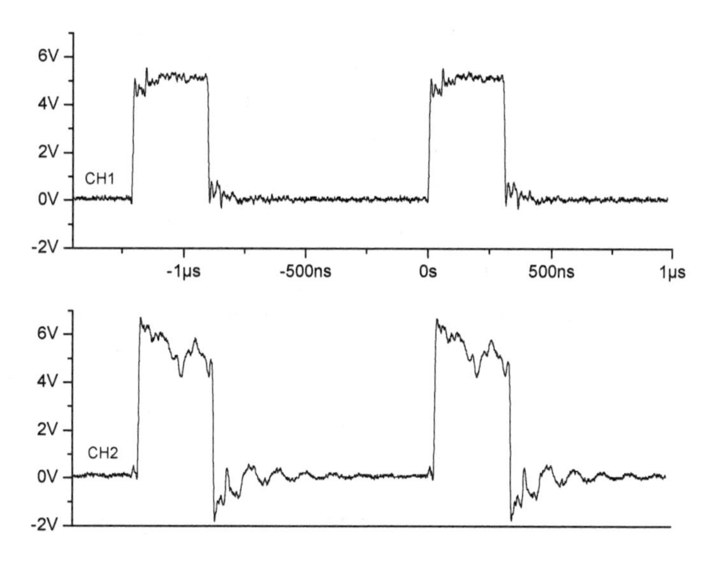

Figure 5.9 Results from Experiment 5.4 step 5.

6. Now terminate the line by connecting resistor *RL*. Observe the voltage on oscilloscope channels 1 and 2. The bottom portion of Figure 5.10 (next page) shows an example channel 2 waveform for this setup. For comparison purposes the top portion is a copy of the channel 2 waveform taken from Experiment 5.2 (the terminated case), and the middle portion is a copy of the bottom of Figure 5.9 (the gate connected but without termination case).

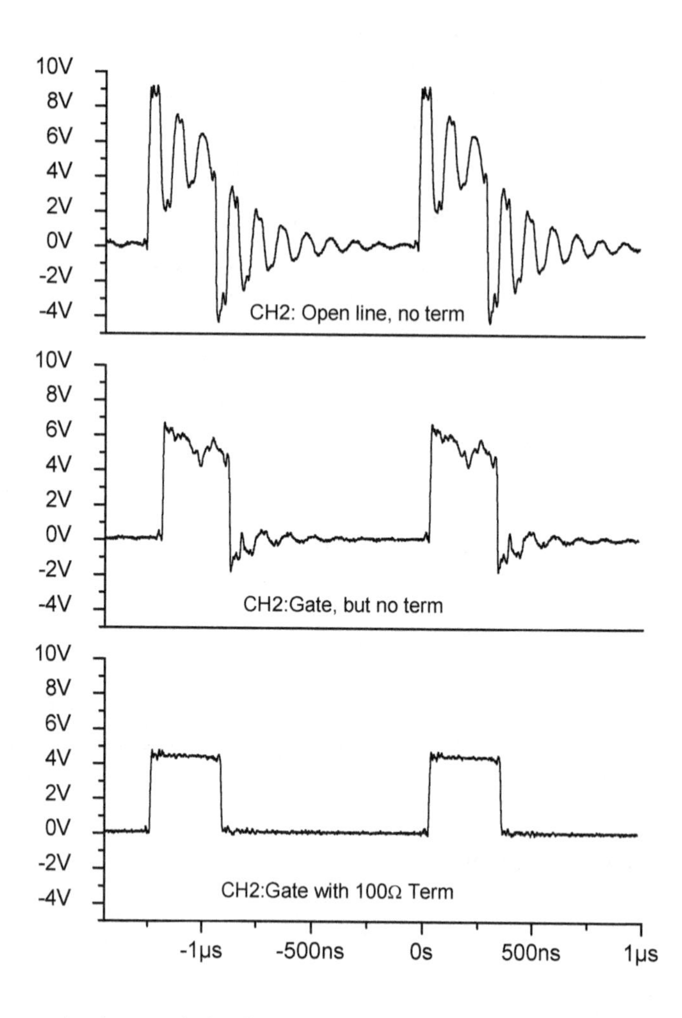

Figure 5.10 Response of an open circuit transmission line (top); diodes within input gate greatly improve response (middle); termination resistor gives best results (bottom). For easy comparison, all three are plotted with the same scale.

Comments

The input resistance measured at pin 2 of a 74HCT240 gate typically is 50meg ohms at room temperature. The input also has a 10pF capacitance, but this value is too low to affect these measurements.

In fact, for the frequencies involved in these experiments, the Pin 2 impedance is essentially that of an open circuit compared to the 100Ω of the transmission line. Because of this it would be reasonable to expect the results from this experiment to look similar to the open circuit results obtained in Experiment 5.2 (Figure 5.4). With this reasoning we'd assume the waveform wouldn't be much different whether pin 2 was connected to the line or not.

However, by comparing the top and middle portions of Figure 5.10 we can see that this is not so. The top portion shows the results from Experiment 5.2 (Figure 5.4), where the end of the line is open circuited (*RL* and pin 2 are not connected). As we saw in that experiment, the waveform shows the excessive ringing and overshooting caused by the large reflections created when the incident wave strikes the open circuit.

The middle portion shows the behavior when Pin 2 is connected and *RL* is not. Although the line still has what amounts to an open circuit at its end the pulse is greatly improved. In fact, it resembles the results shown in Figure 5.7 when a 1N148 diode is used to clamp reflections.

The bottom of the Figure 5.10 shows how adding a 100Ω termination resistor cleans up the waveform even further. In fact, it looks nearly identical to the termination results shown in Figure 5.5 from Experiment 5.2.

Why does connecting pin 2 greatly reduce the reflections, but isn't it as effective as adding resistor *RL*? It turns out that the ESD (Electro Static Discharge) protection circuitry present on pin 2, internal to the integrated circuit, is clamping the waveform in a way similar to the diode clamp explored in Experiment 5.3. The bibliography lists references that discuss ESD circuits, but Figure 5.11 shows a simplified diagram of the ESD protection circuitry commonly found connected to CMOS inputs, including the 74HCT240.

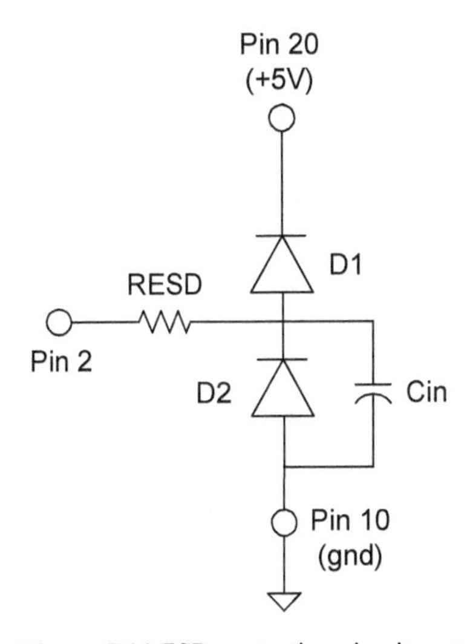

Figure 5.11 ESD protection circuits act as clamp diodes that can limit reflections.

Diode *D1* turns on when the voltage at pin 2 is higher than the voltage on pin 20 (+5V) plus a diode drop (typically 0.6V). Diode *D2* turns on when the pin 2 voltage is more negative than the voltage on pin 10 (ground), minus a diode drop. This means the pin 2 voltage should never exceed +5.6V or -0.6V, but resistor RESD, parasitic resistance of the diodes, noise present on the power and ground lines inside the integrated circuit, and the inductance which is in series with pins 2, 10, and 20 cause the clamping action to occur at voltages outside this range. This is evident in the middle portion of Figure 5.10, where the voltage at pin 2 reaches nearly +7V and -2V. An important but often overlooked point is that RESD makes the actual voltage at the input to the gate inside the integrated circuit lower than what you measure outside, at the pin with an oscilloscope.

Critical Observations

⇒ **The internal circuitry present in CMOS I/O circuits can cause beneficial clamping of the reflections created by the high input impedance of CMOS gates.**

⇒ **The clamping action of the integrated circuit parasitic diodes is not their intended function and they aren't as effective *when measured at the pin* as a good quality clamp you deliberately place on the circuit board (such as those shown in Experiment 5.3).**

Supplemental Exercises

- Use circuit analysis to demonstrate why the clamping action of *D1* and *D2* in this experiment isn't quite as effective as the clamping in Experiment 5.3.
- Review the datasheet for a 74HCT240. How much is the rated current for the ESD diodes? How does this specification compare to the current launched down the cable?

Experiment 5.5: The Effect of Pulse Width on Reflections

Purpose

To determine how reflections and timing can alter the wave shape of a pulse, and how far end termination can be used to improve pulse quality.

- Note 1: This setup is identical to Experiment 5.2, but in this experiment resistor *R1* will be adjusted to change the pulse width and frequency of the transmitted waveform.

- Note 2: The waveform appearing on oscilloscope channel 1 will become distorted as *R1* is adjusted. The signal can become too distorted for some oscilloscopes to use channel 1 as a reliable trigger. If during this experiment your oscilloscope is having difficulty maintaining a solid trigger, use an oscilloscope probe to connect your oscilloscopes "external trigger input" to pin 14 of the 74HCT240, and set it to trigger on the falling edge.

Procedure

1. Follow the build and setup procedures described in Experiment 5.2. Do not yet connect *RL* (it will be connected later, in steps 3 and 4).
2. Apply power and make the output period shorter and longer by adjusting resistor *R1*. Observe the effects this has on the near end (oscilloscope channel 1) and far end (channel 2) waveforms. Experiment with different settings for *R1* and measure the pulse period and width to determine why the waveforms become larger or smaller in amplitude at some settings as compared to other settings.
3. Use a 100Ω resistor for *RL*, and connect it at the far end as is shown in Figure 5.3. Observe how adjustments to *R1*affects the waveforms as compared to other settings. Figure 5.12 (next page) shows an example waveform with and without termination.
4. Repeat step 3 using values for *RL* of 47Ω, and then 220Ω and observe how the waveforms differ from when *RL* was 100Ω.

Comments

The top portion of Figure 5.12 shows the response of the prototype with no termination when *R1*is adjusted such that both the high period low period are about 35ns. The pulse has a repetition period of 70ns, corresponding to a frequency of just over 14MHz.

The cable used in the prototype has a round trip delay of 50ns, so a 35ns wide pulse will have switched from logic high to logic low when the reflection arrives back from the open circuit at the far end. This case can be seen in the "CH1: No termination" portion of the figure.

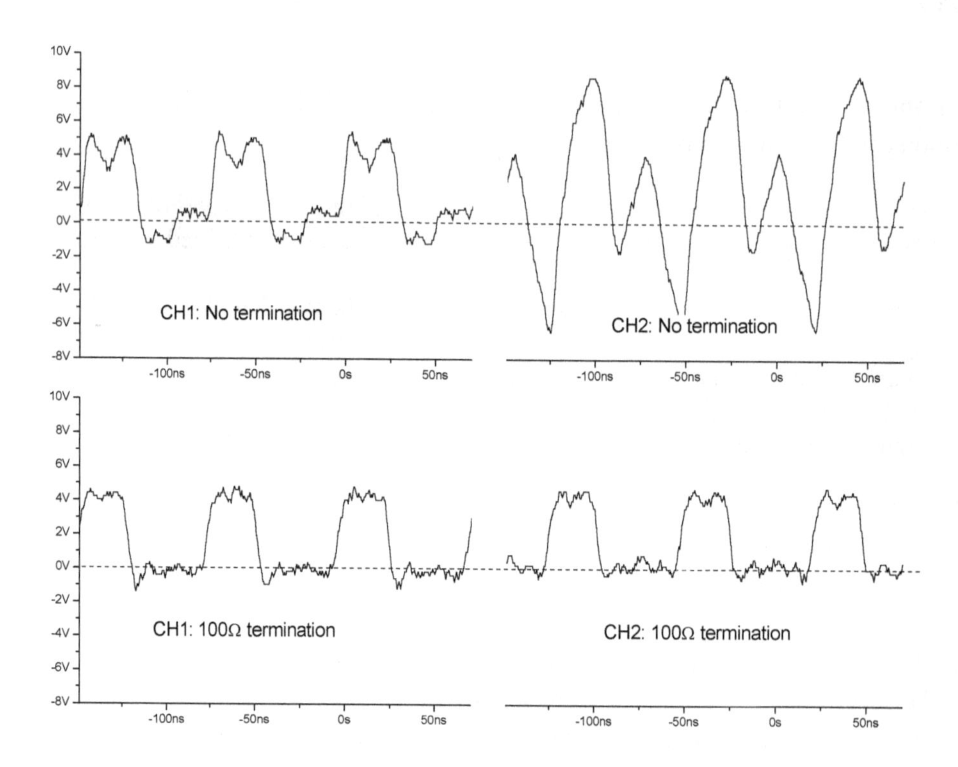

Figure 5.12 Results for Experiment 5.5. Adding termination at the far end greatly improves the signal waveform at both the far and near ends.

The effect of having multiple reflections combine at the far end is very evident in the right portion of the figure ("CH2: No termination"). The voltage peaks (about ±8V) are much higher and very much lower than the launched voltage (about 0 to nearly +5V), and the waveform is significantly distorted.

In fact, various shapes for either of these waveforms can be created by adjusting *R1* to change the relative timing between when a new pulse is created and when a refection arrives from the far end.

The two graphs in the bottom of the figure show the pulses when termination is used to prevent reflections. The signal is free of significant ringing and exhibits little overshooting.

- Note: The horizontal axis in the figure extends from -150ns to +70ns. This is an artifact of the trigger setting: On the oscilloscope used to record these waveforms time 0 occurs at the point of trigger; events stored in the oscilloscope but that took place before the trigger happens occur in "negative time". Times after the trigger occur in "positive time".

Critical Observations

⇒ **When no terminations are present to dampen reflections, it's the timing (in particular, the width and the spacing) rather than the specific frequency of the launched pulses that determines the shape of the received signal appearing on channel 2.**

⇒ **The electrical characteristics of signals greatly distorted by reflections can be significantly improved when the receiving end of the line is terminated in the characteristic impedance of the line.**

Supplemental Exercises

- Create a "bounce diagram" to understand the results of your measurements, especially those that produced the highest voltage on oscilloscope channel 2. Consult the books listed in the Bibliography to find how bounce diagrams (also called "reflection diagrams") are created.

Experiment 5.6: Source Series Termination

Purpose

To observe the way in which source series termination improves the signal waveform at the load, and to determine the proper value of resistance.

- Note 1: This setup is identical to the setup in Experiment 3.5 that was used to determine propagation delay. Before proceeding with this experiment, review experiments 3.2 and 3.5.
- Note 2: You will be repeating this experiment using six different values for *RS*:
  1) 22Ω
  2) 47Ω
  3) 100Ω
  4) 150Ω
  5) 220Ω
  6) 560Ω

  Refer to Appendix A for information regarding suitable resistor types.

Procedure

1. Measure the actual values of the six resistors you'll be using for *RS*. Record their values in your notebook.
2. The test setup is shown in Figure 5.13. For the first iteration of this experiment, use the 22Ω resistor for *RS*.
3. Follow steps 2 through 6 of Experiment 3.5, paying particular attention to the build notes and how to prepare and connect the twisted pair cable. Be sure to adjust resistor *R1* until the pulse at pin 3 of the integrated circuit is at least 200ns wide. Make this adjustment before you connect *RS*.
4. Connect the oscilloscope probes as shown in Figure 5.13, and observe and record the voltages on channels 1 and 2.
5. Switch off the 5V supply, and change *RS* to 47Ω.
6. Switch the 5V supply back on and observe and record the voltages on channels 1 and 2.
7. Repeat steps 5 and 6 for the remaining resistors on your list.
8. Figure 5.14 shows example far end (oscilloscope channel 2) results.

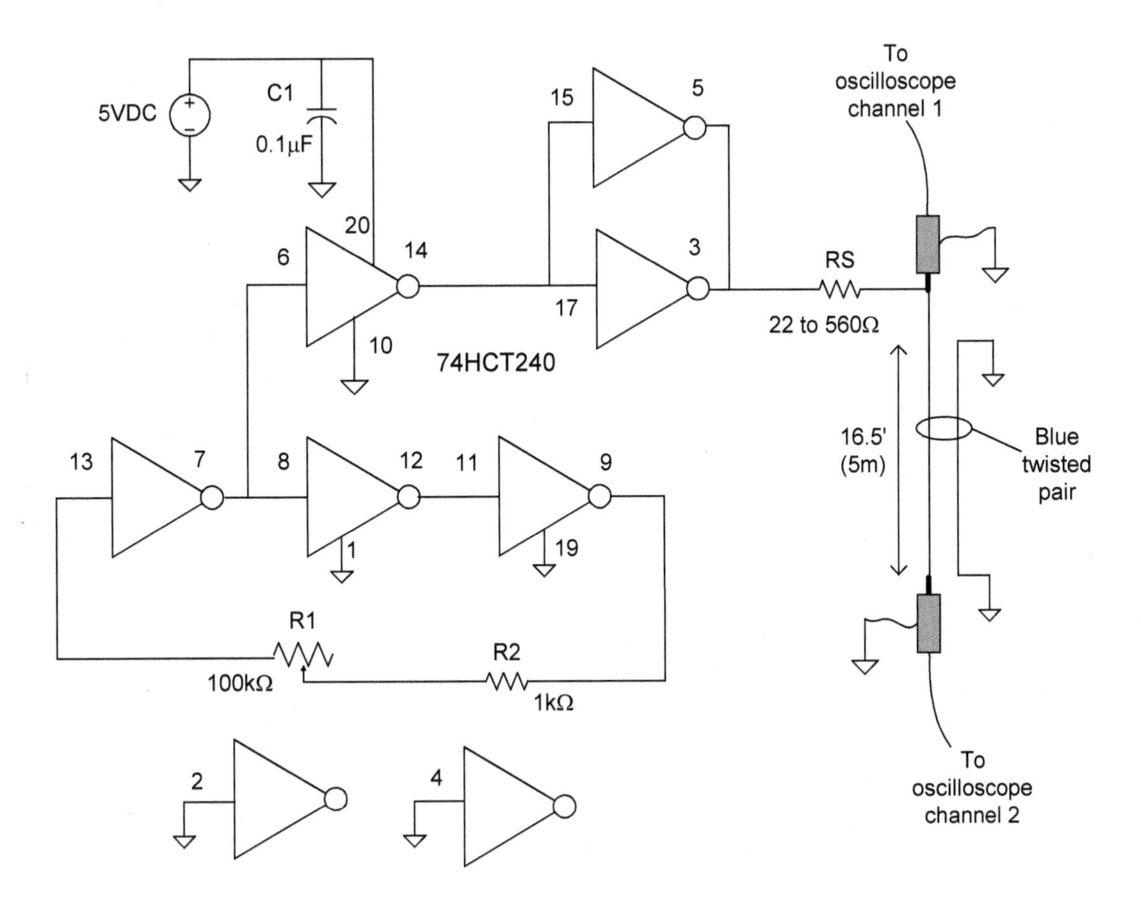

Figure 5.13 Test setup for Experiment 5.6. Resistor RS is varied to show the effects resistance has on signal waveform and reflections.

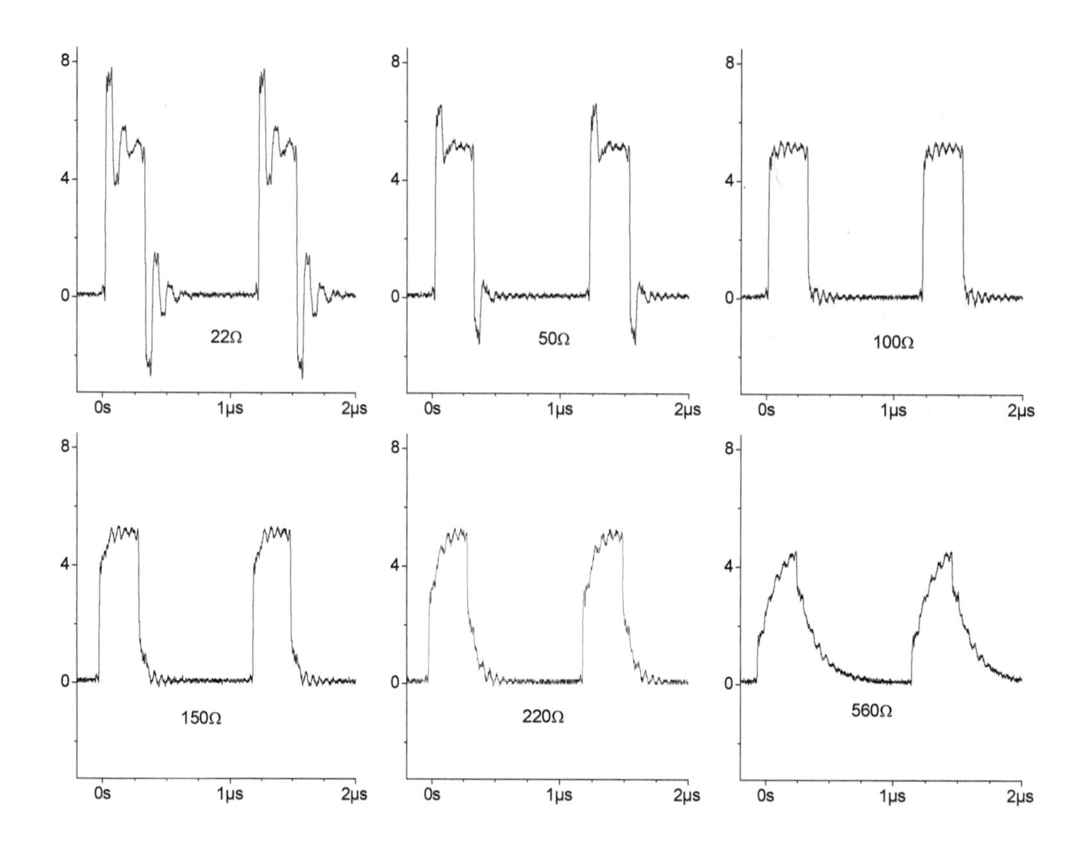

Figure 5.14 Oscilloscope channel 2 (far end) waveforms from Figure 5.13. The cable has an impedance of 100 Ω. Making RS lower than this doesn't remove ringing, but does create sharp edged pulses. Making RS larger than the cable impedance removes ringing but rounds the edge. Very large values create a sawtooth type waveform.

Comments

From previous experiments we know the transmission line impedance is 100Ω, and from the description given in Chapter 4 on source series termination we'd expect making *RS* equal to 100Ω would give the best results. The third panel in Figure 5.14 shows that this is so. Values less than this cause ringing (shown in the figures first two panels), while higher values (shown in the last 3 panels) cause rounding.

Making *RS* less than *Zo* yields sharp edged pulses that can be beneficial in those situations where the receiving logic can't tolerate slow, jittery edges. However, this technique is only effective when the ringing isn't too severe, otherwise the ringing may be interpreted by the receiver as multiple pulses. The ringing also is a source of high frequency noise, which can complicate power supply decoupling and aggravate EMI (electromagnetic interference) problems.

Ringing does not occur when *RS* is greater than *Zo*, but moderately high values causes rounding of the pulse "tail", and can create jagged regions as the reflection travels up and down the line multiple times. Very high values of *RS* result in a saw tooth type waveform because rather than appearing as a distributed capacitance the transmission line capacitance acts as a single lump which is charged exponentially. This rounding can become so excessive that the pulse forms a peak rather than having a flat portion. The 560Ω case in the figure is an example of this.

In general it's best to avoid making *RS* so large that the pulse becomes excessively rounded. A pulse with such poor rise and fall times can make the receiving logic susceptible to noise related mis-triggering.

It would be fair to conclude from the figure that in this setup resistance values ranging between 50Ω and 150Ω would be acceptable, but this judgment very much depends on the application. Some high performance systems (including, but not only, differential signaling) require *RS* to be closely matched to the actual characteristic impedance of the line.

Critical Observations

⇒ **Best signal quality will be obtained when the resistance value for *RS* equals the transmission lines characteristic impedance. However, this isn't always a requirement, and in some applications it's possible to obtain satisfactory results when *RS* is close to the value of *Zo*.**

⇒ **If moderate ringing is acceptable in a particular application, setting *RS* < *Zo* can reduce the time the signal requires to transition through the receivers switch point and so improve switching speed.**

Supplemental Exercises

- Repeat this experiment with pin 2 of the integrated circuit connected at the far end, as was done in Experiment 5.4 (but be sure not to connect load resistor *RL*). How does this alter the results? How critical is the value of *RS* when pin 2 is connected, as opposed to when pin 2 is not connected?
- Experiment 3.4 showed that the transmission line impedance depends on the switching behavior of neighboring wires. Repeat that experiment, but use 100Ω for *RS*. How does the far end waveform appear now, as compared to the 100Ω test performed in this experiment (where *Zo* is known to be 100Ω)?

- Review the waveforms appearing on oscilloscope channel 1 (the near end waveforms) for the 6 different values of *RS*. How does the near end waveform change as *RS* is changed? Why does this happen?

Chapter 6 Fundamentals of Crosstalk

The voltage and current waves travelling down a transmission line can couple unwanted electrical energy which appears as noise to a nearby conductor. Circuit board traces can couple to other traces, to planes within the circuit board, from planes to traces, or from planes to wires lying on top of the board. Wires in cable experience coupling from other wires in the bundle, or from conductors outside the bundle.

The wire or trace carrying a signal is referred to as the aggressor; the wire or trace that receives the unwanted coupled energy is the victim. If large enough, crosstalk voltage pulses can cause victim logic gates to switch when they should not. This can occur even when the gate driving the victim wire isn't switching and is held at a fixed logic level. Crosstalk can also prevent signals that do switch from switching at the proper time, resulting in timing errors (including jitter).

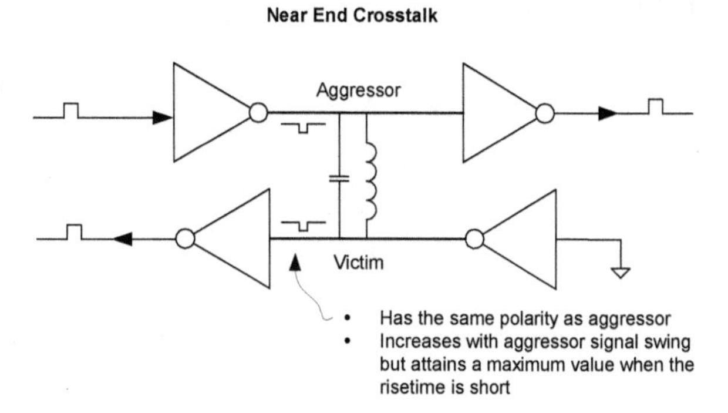

As we'll see in the following sections, crosstalk can be created at either end of the victim wire, and from Figure 6.1 we can see that the two types of crosstalk have very different electrical characteristics.

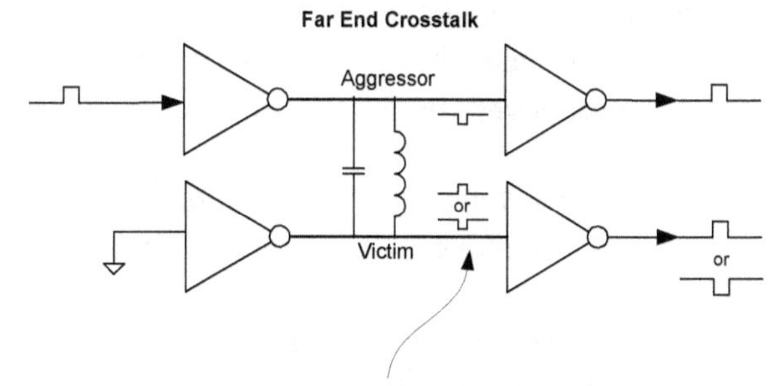

Figure 6.1 Coupling from an aggressor to a victim causes near end and far end crosstalk pulses even though the victim is held at a static logic level.

The nature and behavior of the coupling between the aggressor and victim conductors can be surprising, which is why it can be difficult to understand the root cause of crosstalk problems. This chapter presents some of the underlying concepts necessary for you to understand how crosstalk is created and how it behaves. You'll see in the next chapter's experiments how applying a few basic strategies can often cure all but the most stubborn crosstalk case (or prevent it from occurring in the first place).

Coupling Capacitance and Inductance

Capacitance and inductance is present between circuit board traces routed side by side, or wires in a cable. These are sometimes called parasitics, but in fact the capacitance and inductance are a natural result of the proximity of the two conductors, and (at least for the capacitance) of the insulating material that separates them. For instance, Figure 6.2 represents an end view of two wires in a cable. The wires form the plates of a capacitor, and the insulation surrounding each wire is the capacitors dielectric. The coupling capacitance would be lower if the wires were separated by only air because the *dielectric constant* of air is 2 – 4 times lower than that of the plastic insulation. You may recall from elementary electric field theory that the dielectric constant describes the relative difference in capacitance a structure would have if a dielectric other than a vacuum is used when creating a capacitor. This means capacitors using a plastic with a dielectric constant of 4 would have four times the capacitance of identical capacitors that used only air (which has a dielectric constant very nearly that of 1) as an insulator.

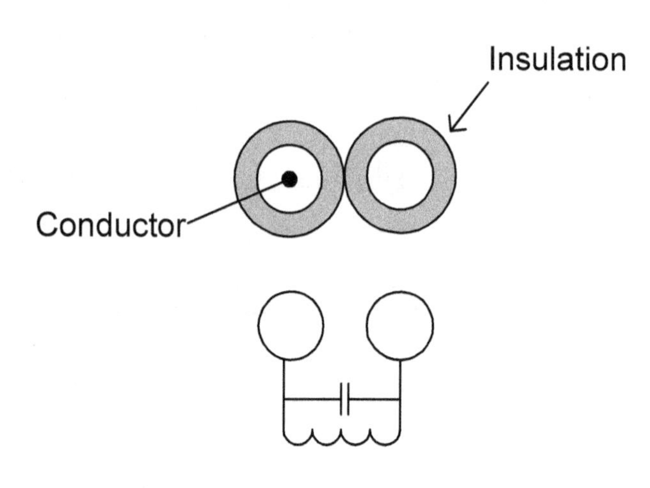

Figure 6.2 Coupling capacitance and inductance is present between two wires.

Separating the wires increases the air gap between the conductors. This lowers the coupling capacitance and so reduces the current the capacitor can inject into the aggressor.

Mutual inductance (inductance appearing between wires) is also present and is a second way the aggressor disturbs the victim. Increasing the separation between wires reduces the mutual inductance between them, and so reduces the ability of the aggressor to magnetically disturb the victim. Unlike with capacitance, the inductance doesn't change if the wires are separated by air or plastic.

The bibliography lists references describing crosstalk in more detail, but this abbreviated explanation of the coupling between conductors gives you a first insight as how to combat crosstalk: Increase the separation between the aggressor and victim conductors. By reducing coupling this attacks the root cause of the problem, and is completely effective if the separation can be made large enough. However, usually the need to route traces close to one another on a densely packed circuit board, or the close, tightly bundled wires in a cable make it impractical to fully correct crosstalk problems with this method. Instead, the termination techniques described later in this chapter, and demonstrated with the experiments performed in the following chapter, are used.

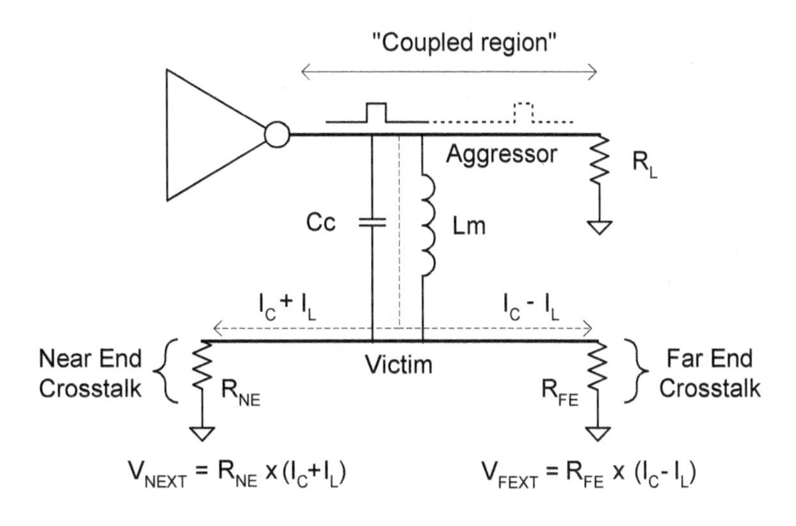

Figure 6.3 Current injected from aggressor to victim splits to create near end and far end crosstalk.

Mutual Impedance

The coupling inductors and capacitors are smoothly distributed all along the length of the conductor, just as we saw in Chapter 2 when discussing impedance and delay. In that case the inductors and capacitors were formed with respect to ground (actually, the signals return path) and so gave rise to the transmission lines characteristic impedance. With coupling, the inductance and capacitance are formed with respect to some other signal conductor and create a *mutual impedance*. Although we won't be discussing mutual impedance any further in this manual, you should know the term because it's used in signal integrity analysis and some CAD tools to predict the amount of crosstalk voltage one line will couple onto another.

Two Types of Crosstalk

Figure 6.3 shows how the coupling capacitance and inductance cause two currents to flow in the victim wire when a pulse travels down the aggressor wire. One of the victim currents flows in the same direction as the aggressor pulse, toward the load, and a second flows in the opposite direction, back toward the source. As you can see, crosstalk voltages are created when the currents flow through load resistors, but (although not shown) crosstalk voltages are also created when the current charges a load capacitor or stores energy magnetically in an inductor at the load.

The voltage created on the victim wire by the currents flowing back toward the source (the near end) is called Near End Crosstalk, which is usually abbreviated as NEXT. The crosstalk voltage created at the load end of the victim wire (the far end), is called Far End Cross Talk. It's abbreviated as FEXT.

The figure shows how in NEXT the current injected by the coupling capacitance (I_c) adds with the current from the mutual inductance (I_L), but in FEXT the inductive current subtracts from the capacitive part. This means NEXT will always be created, but with FEXT it's possible for the inductive coupling to cancel the capacitive coupling and produce no crosstalk. This cancellation happens on circuit board stripline traces that aren't too lossy, which is why signal integrity engineers sometimes say that there is no far end crosstalk on circuit board traces that are routed between two plains. This cancellation doesn't happen with microstrip traces, which is why FEXT is always possible on traces routed on the circuit board's top or bottom surfaces.

If the capacitive part of the FEXT current is larger than the inductive part the microstrip FEXT voltage will be positive (it will have the same polarity as the aggressor pulse). It'll be negative (have a polarity opposite of the aggressor) when the inductive part is greater. Most often the inductance dominates on circuit board microstrips, which makes FEXT negative, but FEXT can become positive if the microstrips are wide or it they are covered in a very thick layer of soldermask or conformal coating because this causes the capacitance to dominate.

The FEXT Pulse

In practice crosstalk is usually a concern when the coupled voltage is high enough to erode input noise margins and so cause the victim logic gates to change state inappropriately ("glitch"), or when it's high enough to affect timing. You'll see in the next section that the FEXT voltage will be highest when the "coupled region" (the distance where the aggressor and victim run next to each other) is long, and when the aggressor pulse has a short rise time and a large change in voltage. You'll also see that the FEXT pulse can have the same or opposite polarity of the aggressor pulse.

A very narrow FEXT pulse may not have enough amplitude or width (time) to falsely trigger a receiving logic gate. In fact, often the worst FEXT offenders are wide, moderately high voltage pulses, and generally the victim logic won't respond to a narrow, higher voltage pulse. As you can see in the following figure, FEXT pulse will be as wide as the aggressor rise or fall time, and the width isn't affected by the length of the line. In fact, a short line will produce the same width FEXT pulse as a long line.

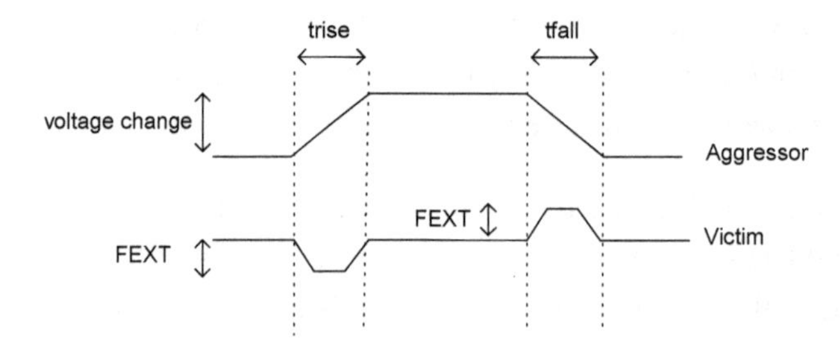

Figure 6.4 FEXT pulse is as wide as the aggressor rise time and occurs on both edges. The pulse may have the same polarity as the aggressor, or the opposite polarity, as is shown here.

These things mean a fast rise time aggressor pulse having a large voltage swing will produce a narrow, high voltage FEXT pulse. An aggressor with a slow rise time having a low voltage swing will produce a wide, low voltage FEXT pulse.

To eliminate FEXT in practice usually means keeping the coupled region short because, as we are about to see, the FEXT pulse amplitude depends on the length of the coupled region. FEXT can also be reduced or eliminated by selecting logic gates that have the slowest possible rise time and lowest output amplitude that will work in your application. In next chapter's experiments you'll see how the reduced voltage swing produced by series termination can help reduce FEXT for this reason.

A Worked Example: The Far End Crosstalk Equation

Equation (6.1) shows how the length of the coupled region (*length*), the aggressor rise time (*risetime*), and voltage swing (*voltagechange*) determine the amount of FEXT voltage.

$$FEXT = Kf \times length \times \frac{voltagechange}{risetime} \tag{6.1}$$

Kf is the "forward crosstalk constant"; its value depends on the culprit and victim impedance, and on how far apart they are from each other. As you might expect, *Kf* is worse when the traces are close together, but it's also larger when their impedance is high. This is why low impedance traces and wires experience lower FEXT than higher impedance ones.

To get the value of *Kf* for a particular situation you'll need to use a field solver or other CAD tool that determines the mutual capacitance and inductance. You then calculate FEXT from those values. We won't be going into those details in this manual, but see the references in the bibliography to learn how to make these calculations. The purpose of presenting equation (6.1) here is to show you with typical numbers how length, the aggressor rise time and change in voltage interact to create a FEXT voltage.

For instance, if we take -170ps/m as a typical value for *Kf* of two closely spaced microstrip circuit board traces, and assume the traces run parallel to each other for 6 inches (0.15 meter), and further assume the aggressor pulse switches from 0V to 3.3V in 10ns, we find:

Kf = -170ps/m
length = 0.15m
voltagechange = 3.3V
risetime = 10ns

$$FEXT = -170ps/m \times 0.15m \times \frac{3.3V}{10ns} = -8mV$$

The minus sign means the FEXT pulse's polarity is the opposite of the aggressor polarity. It will go low (below ground, if necessary) by 8mV on the rising edge of the aggressor, and will go high by 8mV on the falling edge. This is shown below, in Figure 6.5. If the equation had produced a positive value the FEXT pulse would go high on the rising edge of the aggressor and low on the falling edge.

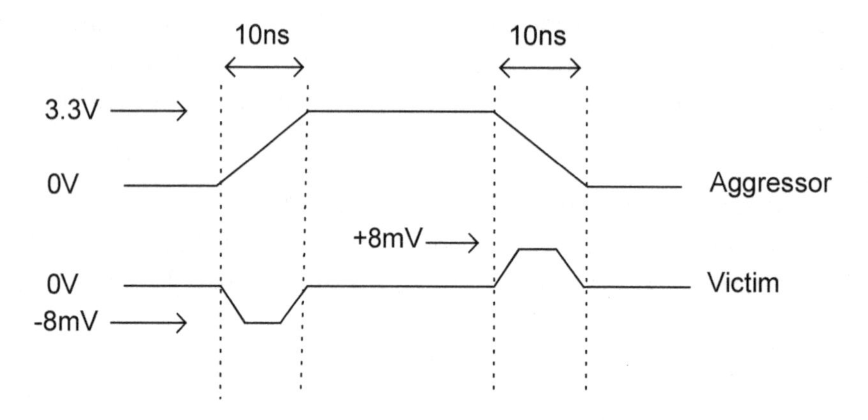

Figure 6.5 FEXT worked example. The FEXT pulse has the opposite polarity from the aggressor because in this example *Kf* is negative.

A FEXT pulse is created every time the aggressor switches. Because in this example *Kf* was negative the FEXT pulse in the figure becomes -8mV on the aggressor's rising edge and +8mV on the aggressors falling edge. *Voltgechange* is made -3.3V in the equation to signify a high to low going transition.

In practice integrated circuit drivers rise and fall times differ, which means the FEXT pulse won't be as symmetrical as the figure suggests. Often the aggressor's falling edge is sharper (faster) than the rising edge, which in the above example would make the -8mV part of the FEXT pulse wider than the +8mV part. This would also increase the +8mV to a higher value.

Notice the equation uses the change in voltage rather than the value of the voltage. This means an aggressor that switches from 12V to 15.3V would cause the same ±8mV FEXT voltage as the 0 to 3.3V signal used in this example because in both instances *voltagechange* is 3.3V.

Increasing the rise time (making it slower) reduces FEXT, and this is one of the most common ways signal integrity engineers fix FEXT problems. For instance, increasing the rise time from 10ns to 20ns cuts the FEXT voltage in half to -4mV. As you'll see, NEXT doesn't behave in this way.

A Word About *Kf*'s Units

From the above example we can see that *Kf* has units of time per distance (such as ps/m). By doing a dimensional analysis we can see why this strange set of units makes sense: the time in *Kf* cancels with the time in *risetime*, while the meter part of *Kf* cancels with *length*. The result is volts.

The NEXT Pulse

The width and voltage (height) of the NEXT pulse depends on how long the coupled region is, and by the value of the *reverse crosstalk coefficient* (*Kb*).With NEXT the length is specified as a time with units such as ps or ns. This is different from what we just saw with FEXT, where there the coupled distance is specified as a distance with units such as inches or cm.

A NEXT "short line case" occurs when the coupled region is shorter than half the rise time of the aggressor. For instance, a 1ns long circuit board trace is "short" if the aggressor rise time is 2ns or more. Circuit board traces often fall into this category. With short lines the NEXT amplitude only depends on the rise time and the aggressor's change in voltage:

$$NEXT_{short} = Kb \times voltagechange \qquad \text{(used when } td < 0.5\text{X rise time)} \qquad (6.2)$$

A "long line" case occurs when the coupled region is longer than half the rise time of the aggressors signal. A 5ns long transmission line is "long" if the aggressor rise time is 10ns or less. Cables often fall into this category, as do circuit board traces carrying very high speed signals. Long lines produce the widest and tallest NEXT voltage, and in fact, long line NEXT will always be greater than short line NEXT. In long line situations the width of the NEXT pulse will be twice the time of the coupled region (which must be specified in terms of time). The amplitude of long line NEXT is proportional to the length of the coupled region and to the ratio of the voltage change to the rise time of the aggressor pulse:

$$NEXT_{long} = 2Kb \times td \times \frac{voltagechange}{risetime} \qquad \text{(used when } td > 0.5\text{X rise time)} \qquad (6.3)$$

In these equations *Kb* is the "backwards crosstalk constant", *td* is the length (measured in time) of the coupled region, *voltagechange* is the aggressor signal swing, and *risetime* is the aggressor rise time. To calculate *Kb* you'll need to use an electromagnetic simulator to determine the inductance and capacitance of your transmission line. The worked example later in this chapter gives a typical value for circuit board traces. *Kb* will be lowest when the impedance is low, and when the spacing between the aggressor and victim traces is large.

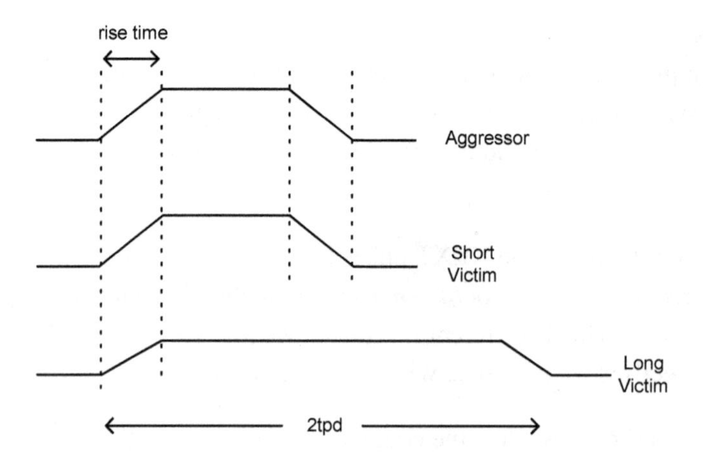

Figure 6.6 NEXT pulse width equals two times the length of the coupled region when the line is long. It equals the aggressor pulse width when the line isn't long.

A Word About Kb's Units

Kb is unitless, which can be seen by performing a dimensional analysis on equation (6.3). The time unit in *td* cancels with the time units in *risetime*, leaving only volts as the result.

What Trace or Wire Lengths are Considered "Long"?

Before deciding on which NEXT equation to use you must first express the distance of the coupled region in time (such as ps or ns) rather than in inches or centimeters (as when you calculate FEXT).

Chapter 2 shows you how to convert a length into a time delay, and by comparing various delays with various rise times it's possible to see if the NEXT will exhibit long line or short line behavior. This has been done in Table 6.1. We've assumed the delay for circuit board traces is 170ps/inch (68ps/cm) because that is a good compromise between microstrips (which generally is a bit faster than this) and striplines (which is generally somewhat slower). The delay for cable is assumed to be 120ps/inch (48ps/cm), which is typical for many twisted pair cables and wire bundles.

Table 6.1 NEXT pulses traveling down wires longer than shown will behave as "long line" crosstalk.

Aggressor Rise Time ns	FR4 Traces Inch (cm)	Wire Cable Inch (cm)
0.25	0.75(1.9)	1 (2.6)
0.5	1.5 (3.8)	2 (5.0)
1	3.0 (7.6)	4 (10)
5	15 (38)	20 (50)
10	30 (76)	40 (100)
20	60 (150)	80 (200)

From the table we see that a victim trace on FR4 that's coupled to an aggressor for a distance of 3 inches (7.6cm) or greater would experience long line NEXT behavior if the aggressor's rise time was 1ns or less.

Use the table as a guide in deciding if the NEXT in your situation will be long line or short line in nature, but don't use it as an indicator of the crosstalk's severity. Use the crosstalk equations (or a circuit simulator) instead.

A Worked Example: The Near End Crosstalk Equation

In practice you'd use a field solver to determine the values of the coupling capacitance and inductance, and calculate *Kb* with them. The references in the bibliography show how this is done. A reasonable value for *Kb* is 0.08 for FR4 circuit boards having closely spaced 60Ω traces. It will be larger for higher impedance traces, and smaller for lower impedance ones, or when the traces are further apart.

We'll determine the NEXT voltage of the same aggressor/victim pair we analyzed previously for FEXT. From that example we know the rise time is 10ns and the change in voltage is 3.3V. The length of the coupled region is 0.15m, but the NEXT equations require the length (*td*) in terms of time. We said previously that a delay of 170ps/inch (6.7ns/m) is a reasonable assumption for common circuit boards. Using this, $td = 6.7ns/m \times 0.15m = 1ns$.

This is a "short line" NEXT situation because *td* is 1ns, which certainly is less than half of the 10ns rise time. This tells us that NEXT will be lower than it would be if the line were long, and that equation (6.2) is the correct one to use. When the aggressor switches from 0V to 3.3V, the NEXT voltage will have a value of:

$$NEXT_{short} = 0.08 \times 3.3V = 264mV$$

NEXT pulses are created on each edge of the aggressor pulse; the NEXT value will be -264mV when the aggressor switches from 3.3V back to 0V (in this situation *voltagechange* is taken as negative to show a high to low transition).

We previously saw that increasing the aggressor rise time from 10ns to 20ns cut FEXT in half, and we said this technique was a common FEXT remedy. However, when the line is short increasing the rise time won't improve FEXT. The line continues to be "short" even when the signal rise time is lengthened, and equation (6.2) still applies (and gives the same answer as before). This tells us that when the line is short making the rise time longer by choosing logic devices having slower rise times won't fix short line NEXT problems. Short line NEXT can only be corrected by reducing the aggressor's voltage swing or by reducing *Kb* (which is done by increasing the spacing between the aggressor and victim, or lowering its impedance).

However, using a logic gate having a faster rise time completely changes things. For instance, if the rise time is reduced to 0.25ns by using a fast logic device the line appears to be "long" because the coupled length *td* (which we already know is 1ns) is greater than half the rise time of 0.25ns. Equation (6.3) applies in this situation, so now the rise time and the line length do matter. In this situation the NEXT voltage becomes remarkably high:

$$NEXT_{long} = 2 \times 0.08 \times 1ns \times \frac{3.3V}{0.25ns} = 2.1V$$

Because NEXT pulses are created on each edge of the aggressor pulse the NEXT value will be -2.1V when the aggressor switches from 3.3V back to 0V (a logic high to low transition).

Increasing the rise time (making it longer) does improve long line NEXT. It's also improved by decreasing the aggressors signal swing and by reducing *Kb*.

Long and Short Line NEXT Summary

- A NEXT pulse is created when the aggressor voltage changes: NEXT pulses occur on both the rising and falling edge of the aggressor pulse.
- NEXT will be the highest when the line is long. It can be reduced by reducing *Kb* (which means increasing the separation between the aggressor and the culprit), making the coupled region shorter in length, reducing the voltage swing of the aggressor, and increasing the signal rise time (making it slower).

- When the line is short the only way to lower the NEXT voltage is to either reduce *Kb* (increasing the aggressor and victim separation), or by reducing the voltage swing of the aggressor. Increasing its rise time won't reduce NEXT in this case.

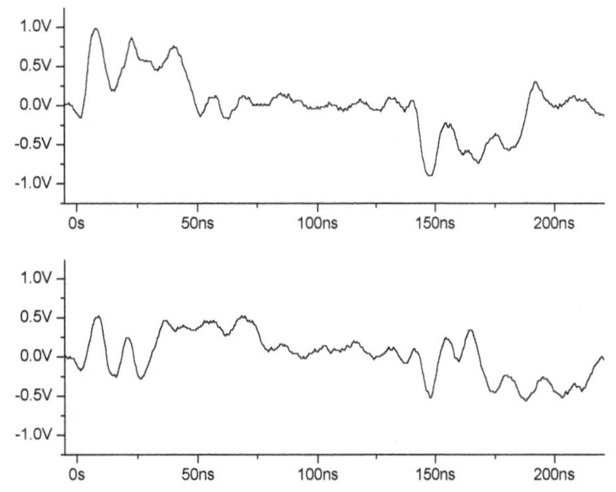

Figure 6.7 Measured FEXT (top) and NEXT (bottom) on actual hardware often don't resemble the ideal waveforms presented in textbooks.

The Importance of Termination

The NEXT and FEXT you measure in your system probably won't look much like what's shown in Figures 6.5 and 6.6: They are more likely to look like the ones in Figure 6.7.

Reflections have caused the shapes to become distorted versions of the ideal ones shown in the previous discussions, and they make the amplitudes not match the values calculated by the equations.

This distortion happens because reflections on the aggressor wire act as signal pulses that couple added energy into the victim wire, creating additional victim crosstalk. Until now our discussion of crosstalk has tacitly assumed reflections were not present, but in fact the coupled waveform can become very complex if multiple aggressor reflections are created. This is especially likely to occur unless both ends of the aggressor are terminated.

In a similar way, reflections will be present on the victim wire unless it's terminated at both ends. The effects of extra coupled energy from reflections occurring along the aggressor wire combining with reflections on the victim wire can create waveforms that hardly resemble the ideal waveforms shown in the previous sections. In fact, the crosstalk you measure in actual systems (or that you model with a circuit simulator) will closely match what's shown in the figures and predicted by the equations only when both the aggressor and victim lines are properly terminated so that no reflections are present on either line.

Chapter 7 Laboratory Exercises: Measuring Crosstalk

Normally designers want to prevent or minimize crosstalk, but in these experiments you'll intentionally create crosstalk and measure its effects. You'll see how different types of termination schemes can be used to reduce or eliminate crosstalk.

Experiment 7.1: Crosstalk Under Ideal Conditions

Purpose

To observe how closely crosstalk in a properly terminated set of transmission lines matches the ideal description given in Chapter 6.

Procedure

1. The test setup is shown in Figure 7.1. If you have not already performed any of the experiments in Chapters 3 or 5, before proceeding with this experiment check the build notes appearing in Appendix A for construction details and circuit operation, and for information regarding parts substitution.
2. If you have not already done so, prepare 16.5 feet (5 meters) of cable as described in Appendix B.
3. If you've not already done so as part of Experiment 3.3, untwist the orange pair at one end so they become two separate wires. One of the wires will be colored all orange; the other will be colored white with an orange stripe. The brown pair will not be used in this experiment. The length where the wires are untwisted isn't critical but it shouldn't exceed 5 inches (13cm). Remove about ¼ inch (0.6cm) of insulation from the orange wire. The white wire with the orange stripe isn't used and won't be connected.

4. Connect the three wires (the blue pair, and the single orange wire) as shown in the figure.

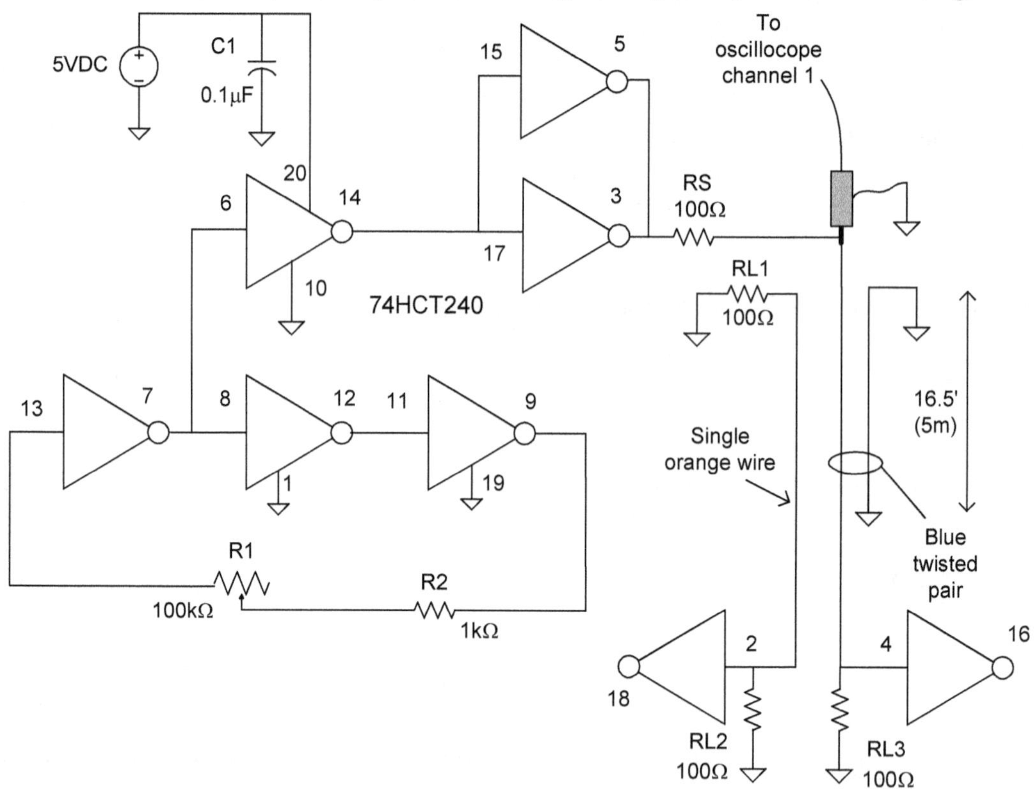

Figure 7.1 Test setup for Experiment 7.1.

5. Connect the channel 1 probe as shown. Set the oscilloscope to trigger on the rising edge of channel 1.
    a. Note: Short ground leads are required on the probe to obtain the best results with this experiment. See Appendix C for details.
6. Apply power to the circuit and adjust resistor *R1* so that the pulse appearing on channel 1 is at least 200ns wide. The actual width isn't critical, and provided the width is more than 200ns there's no need to make a change if you've already set this up in previous experiments.
7. Use the channel 2 oscilloscope probe to alternately observe the near end (NEXT) and far end (FEXT) crosstalk voltages. For best results use short ground leads on the probe.
    a. NEXT is measured at *RL1*
    b. FEXT is measured at *RL2* (pin 2 of the integrated circuit)
8. Sample waveforms appear in Figure 7.2. The waveform marked "Aggressor" is the signal observed on oscilloscope channel 1.

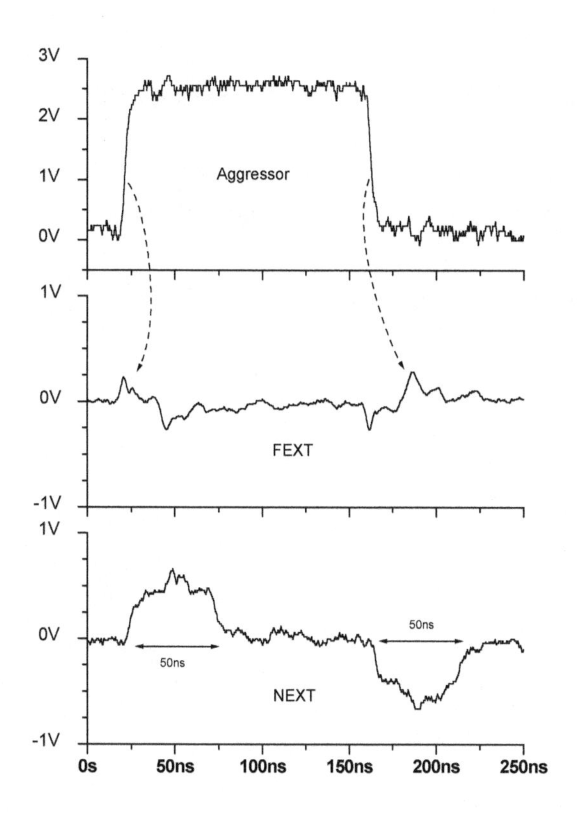

Figure 7.2 Results from Experiment 7.1. FEXT spikes and NEXT pulses are created on the rising and falling edge of the aggressor.

Comments

The width of the FEXT pulse is roughly equal to aggressor's rise time, and the FEXT pulse occurs on both the rising and falling edges.

From previous experiments we know the delay time is about 25ns for the particular cable being used in in these measurements. The aggressor rise time is about 10ns, so from Chapter 6 we know this transmission line is in the "long line" category. This means the NEXT pulse width should be 2 times the delay (i.e., 50ns) in this setup, and that is in fact what the figure shows.

Even though these wires are terminated, probe inductance, and circuit wiring inductance distort the pulses so they aren't exact matches of the idealized pulses described in Chapter 6.

Supplemental Exercises

- Use a long ground lead on the channel 2 probe. How does this mask the true nature of the crosstalk voltages?
- Disconnect pins 2 and 4 from *RL2* and *RL3*, and then connect pins 2 and 4 to ground. Measure FEXT and NEXT at *RL2* and *RL1*. Are the NEXT and FEXT waveforms different? Why?
- Vary the aggressor frequency by adjusting resistor *R1*. How does this affect the crosstalk?

Experiment 7.2: Termination Effects on Far End Crosstalk

Purpose

To observe the effects terminating the aggressor and victim lines has on far end crosstalk.

Background Discussion

In this experiment a driver sends a signal down a long transmission line to a receiver. The input of a victim receiver is connected to ground through a wire in the same cable bundle with the expectation that the victim receiver will not switch. Various termination schemes will be used in an attempt to improve crosstalk.

To obtain a baseline result, perform Experiment 7.1 before attempting this experiment

- Note: Use a dual channel oscilloscope that has an external trigger input. This experiment assumes probes are connected to oscilloscope channel 1, channel 2, and to the external trigger input.

Procedure

1. The test setup is shown in Figure 7.3. It is identical to the setup used in Experiment 7.1, with the exception that resistor *RL1* is replaced with a short circuit, and resistor *RL3* is removed.
2. Do not yet connect resistor *RL2*. It will be connected later, in step 7.
3. Apply a short jumper wire across resistor *RS*, shorting it out. This connects integrated circuit pins 3 and 5 directly to the blue wire and represents the case where the aggressor is unterminated. The jumper will be removed in a later step so you can see how adding *RS* changes crosstalk.

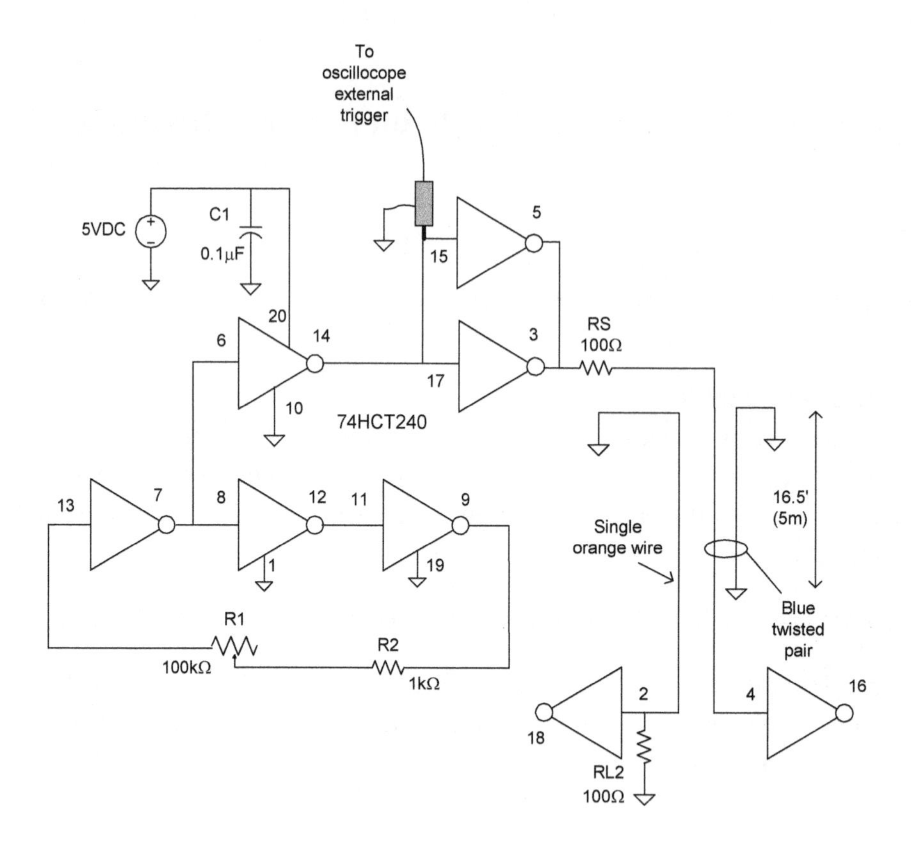

Figure 7.3 Hardware test setup for Experiment 7.2.

4. Connect the probe that's attached to the oscilloscopes external trigger input as shown in the figure.
    a. Set the oscilloscope to trigger on the falling edge of the external trigger input.
5. Probe pin 3 of the integrated circuit with oscilloscope channel 1. Apply power to the circuit and adjust resistor *R1* so that the pulse is at least 200ns wide. The actual width isn't critical.
6. Probe integrated circuit pin 2 with oscilloscope channel 1, and pin 18 with oscilloscope channel 2. From a logic design perspective the intent is for pin 18 to remain at a logic high level, but in the prototype the crosstalk was sufficient to cause the pin to pulse low. Figure 7.4 shows an example result.

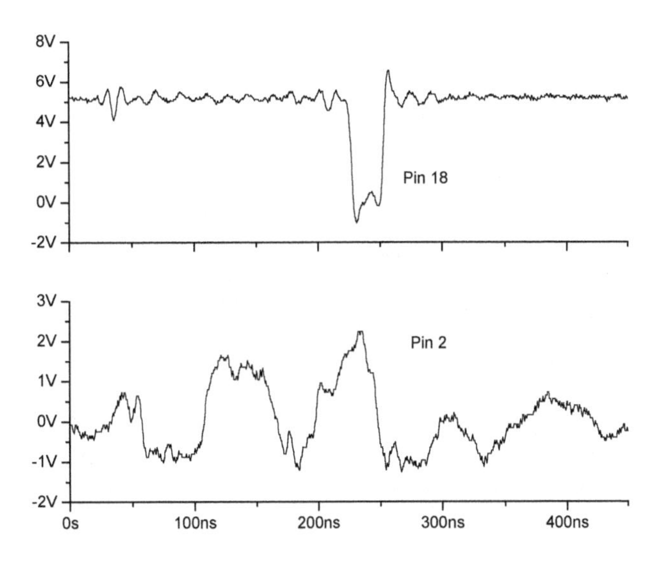

Figure 7.4 Results for Experiment 7.2. When termination resistor RL2 is not present FEXT on inverter input pin 2 is large enough to cause the inverter output (pin 18) to switch.

7. Attach 100Ω resistor *RL2* and repeat the measurements described in step 6. This shows how far end termination at the victim can improve crosstalk.
    - Note: In the prototype hardware, adding *RL2* reduced crosstalk enough to prevent pin 18 from glitching, but this by itself may not be sufficient in your setup.
8. Remove *RL2* and remove the jumper wire that is shorting out *RS*. Repeat the measurements made in step 6. This shows the effects source series termination of the aggressor can have on crosstalk.
    - Note: In the prototype hardware, adding *RS* reduced crosstalk enough to prevent pin 18 from glitching, but this by itself may not be sufficient in your setup.
9. Remove resistor *RL2*, and reattach the jumper wire across resistor *RS*, shorting it out. Replace the short to ground connection at the end or the orange wire with a 100Ω resistor and repeat step 6.
    - Note: In the prototype hardware this change was enough to prevent pin 18 from switching, but this by itself may not be sufficient in your setup.

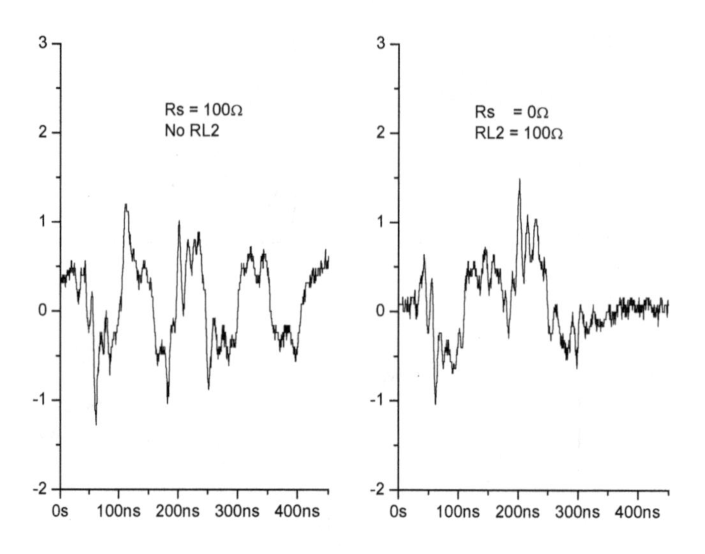

Figure 7.5 Effects of source series terination (left pannel) and far end paralell termination (right pannel) on FEXT. Both termination types reduced FEXT enough to prevent inverter pin 18 from switching.

Comments

The measurements made on the prototype hardware appear in 7.5. The left panel shows the voltage on pin 2 when the aggressor is source terminated, but the victim has no termination (corresponding to experimental step number 8). The right panel shows the pin 2 voltage when the victim is terminated and the aggressor is not (corresponding to step 7). In the prototype hardware pin 18 remains at a logic high for both of these cases and so isn't shown in the figure.

Notice how in both cases termination has reduced the crosstalk voltage and changed its shape from that shown in Figure 7.3 (the unterminated case). Although crosstalk is significantly higher than the crosstalk shown in the properly terminated baseline Experiment 7.1, the terminations used in this experiment are sufficient to keep crosstalk below the point where pin18 is falsely triggered. Nonetheless, only a few hundred millivolts of power supply noise would be needed to cause pin 18 to glitch. Even with termination if your setup is nosier than the prototype you may see faint glitches on pin 18.

Experimental steps 6, 7 and 8 show how crosstalk coupled onto the victim will reflect off a short circuit at the near end and combine to create very noisy waveforms. Step 9 shows that by connecting the far end through a terminating resistor rather than a direct short circuit the reflection is removed, greatly improving the victim waveform.

Critical Observations

⇒ **Proper termination on the victim line can significantly improve crosstalk.**

⇒ **Terminating the aggressor prevents reflections, reducing the noise coupled onto the victim.**

⇒ **Series terminating the aggressor (resistor *RS* in the figure) improves crosstalk in two ways: It reduces the voltage swing and prevents multiple reflections.**

Supplemental Exercises

- Repeat this experiment with different values for *RS* and *RL2*. How sensitive is crosstalk to the value of the termination resistor?
- Use *R1* to adjust the frequency for the aggressor pulse. How does this influence the efficacy of the termination?
- Try this experiment with shorter lengths of cable. How would you expect a cable half the length to affect FEXT?

Experiment 7.3: Termination Effects on Near End Crosstalk

To observe the effects that terminating the aggressor and victim lines has on near end crosstalk.

Background Discussion

This experiment is identical to Experiment 7.2, except that here the victim receiver is placed at the near end, close to the aggressor instead of at the far end.

To obtain baseline results perform Experiments 7.1 and 7.2 before attempting this experiment

- Note: Just as you did in Experiment 7.2, use a dual channel oscilloscope that has an external trigger input. This experiment assumes probes are connected to oscilloscope channel 1, channel 2 and to the external trigger input.

Procedure

1. The test setup is shown in Figure 7.6. It is identical to the setup used in Experiment 7.2, except that the connection to pin 2 of the integrated circuit has been moved from the cables far end to its near end.
2. Do not yet connect resistor *RL2*. It will be connected later, in step 7.
3. Apply a short jumper wire across resistor *RS*, shorting it out. This connects integrated circuit pins 3 and 5 directly to the blue wire and represents the case where the aggressor is unterminated. The jumper will be removed in a later step so you can see how adding *RS* changes crosstalk.

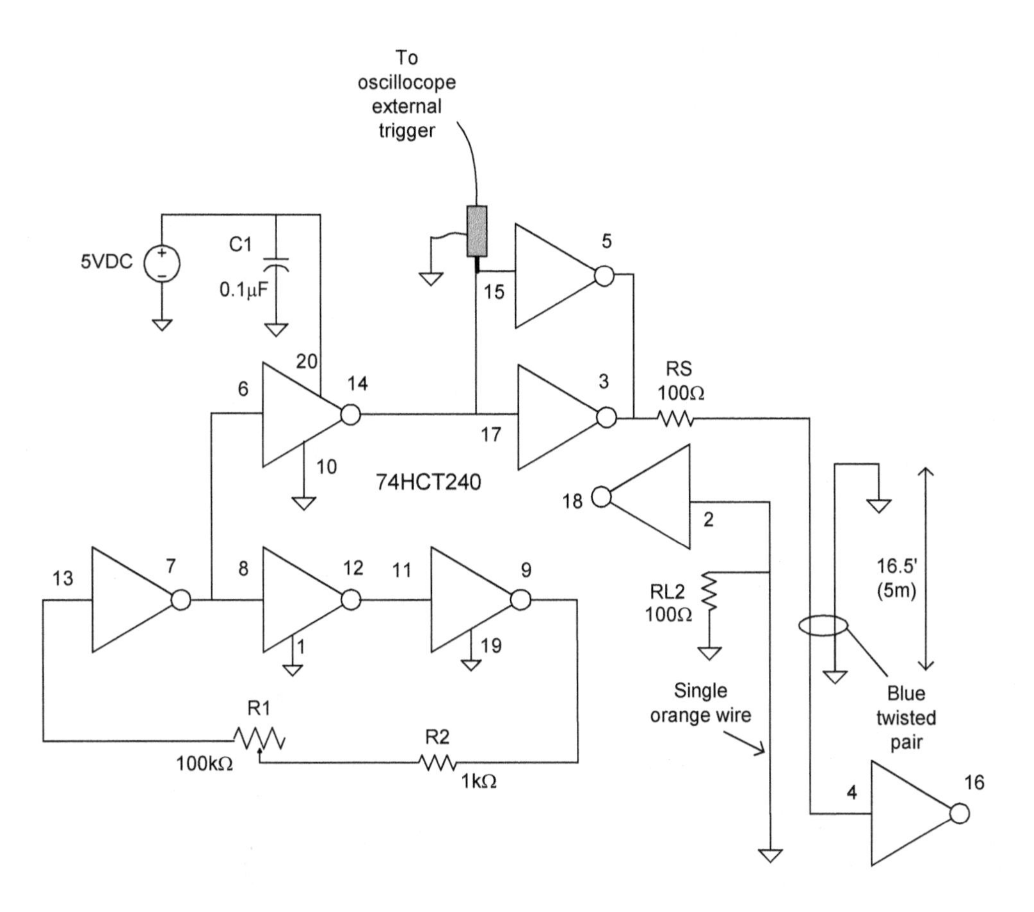

Figure 7.6 Setup for Experiment 7.3 for testing NEXT.

4. Connect the probe that's attached to the oscilloscopes external trigger input to pin 15 as shown in the figure.
    b. Set the oscilloscope to trigger on the falling edge of the external trigger input.
5. Probe pin 3 of the integrated circuit with oscilloscope channel 1. Apply power to the circuit and adjust resistor *R1* so that the pulse is at least 200ns wide. The actual width isn't critical.
6. Probe integrated circuit pin 2 with oscilloscope channel 1, and pin 18 with oscilloscope channel 2. From a logic design perspective the intent is for pin 18 to remain at a logic high level, but, as you can see in Figure 7.7, in the prototype the crosstalk was sufficient to cause pin 18 to pulse low.

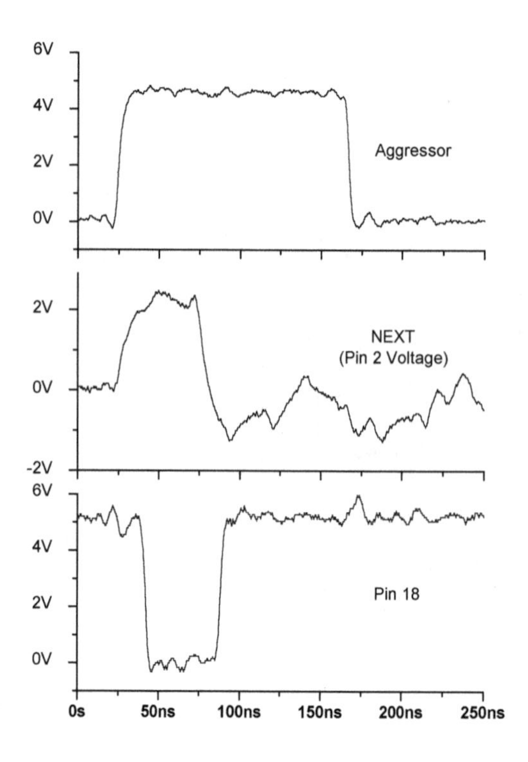

Figure 7.7 Waveforms from Step 6. NEXT appearing on inverter pin 2 is high enough to cause the inverter output (pin 18) to pulse low rather than staying at a logic high as it should.

7. Attach 100Ω resistor *RL2* and repeat the measurements described in step 6. This shows how far end termination at the victim can improve crosstalk. The top portion of Figure 7.8 shows the crosstalk measured on the prototype hardware at pin 2.
    a. Note: In the prototype hardware, adding *RL2* reduced crosstalk enough to prevent pin 18 from glitching low, but this by itself may not be sufficient in your setup.
8. Remove *RL2* and remove the jumper wire that is shorting out *RS*. Repeat the measurements made in step 6. This shows the effects source series termination of the aggressor can have on crosstalk. The bottom portion of Figure 7.8 shows the crosstalk measured on the prototype hardware at pin 2.
    a. Note: In the prototype hardware, adding *RS* reduced crosstalk enough to prevent pin 18 from glitching low, but this by itself may not be sufficient in your setup.

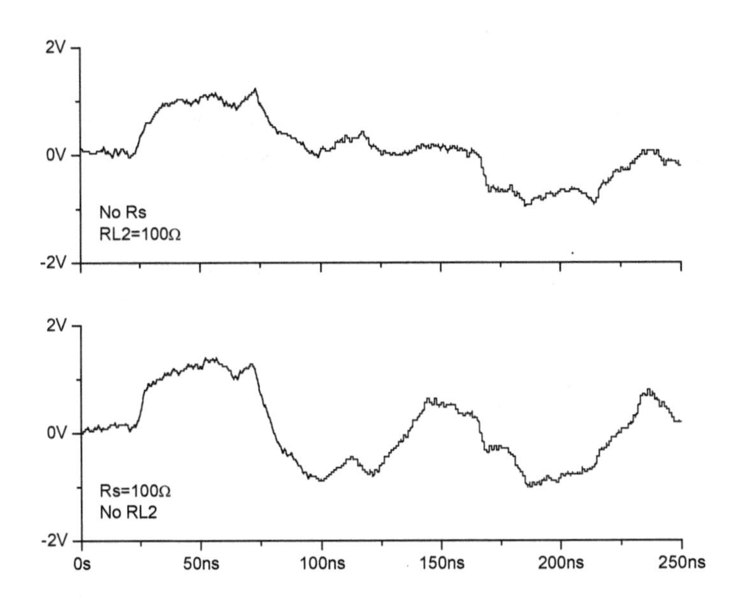

Figure 7.8 Results from Step 7: The pin 2 voltage (NEXT) when the line is source series terminated (bottom panel) and when its parallel terminated at the load end (top panel). Doing either of these things was sufficient in the prototype to reduce NEXT enough to prevent inverter pin 18 from pulsing low.

Comments

Without termination the NEXT voltage is high and wide enough to cause the output of the inverter (pin 18) to glitch low. In fact, as can be seen in Figure 7.7, the pulse is 50ns wide. We know from earlier experiments that this is two times the transmission line delay, which is the width the NEXT pulse should have when the line is "long".

Terminating pin 2 reduces NEXT to the point where (at least in the prototype) the voltage isn't high enough to cause pin 18 to switch low. However, you can see in the top portion of Figure 7.8 that the voltage hasn't been reduced significantly and only a small percentage of additional noise (either power supply noise or additional coupling onto the victim wire, or both) could be enough to cause pin 18 to glitch.

Similarly, you can see in the figures bottom portion that adding series terminator *RS* to the aggressor reduces the NEXT voltage, but not by a great deal. In fact, the amplitude is about the same as the case when *RS* is zero. However, the waveform shape is very different – by adding *RS* the aggressor is properly terminated and additional reflections aren't created along its length. This makes the shape of the NEXT pulse more distinct, and you can now see that it has a width of 50ns (the expected width of the long line NEXT case).

Supplemental Exercises

- Repeat step 6, but adjust *R1* to vary the frequency. How does the NEXT pulse width change as *R1* is adjusted? Why? Does adding *RS* or *RL2* alter your results?
- How does simultaneously adding both *RS* and *RL2* alter NEXT?
- Try this experiment with shorter lengths of cable. How would you expect a cable half the length to affect NEXT?

Appendix A. Test Setup Build Notes

The experiments in this manual use inexpensive parts and are designed to be easy to setup and perform. Alternatives for each of the components are listed later on in this appendix if you wish to make substitutions.

Some of the experiments require you to observe the difference in operation when changing the connection to the cable (which is the transmission line). A small solderless breadboard works well electrically for these experiments and with it you can easily make the necessary changes to the setup. Just be sure to keep the wiring neat and the lead lengths short. Figure A.1 shows how the prototype was wired.

Figure A.1 Prototype wiring on a solderless breadboard. Note short leads and clean layout. Figure A.2 shows the corresponding schematic.

Wiring

The 74HCT240 can oscillate at frequencies of 20MHz and higher, so it's important to keep the wiring neat and the leads short. Notice the prototype wiring of the +5V and ground connections: The wire supplying +5V from the power supply connects directly to pin 20, and the ground from the power supply connects directly to pin 10. These connections aren't supplied by jumpers from the breadboard power buss bars since that type of wiring adds inductance (which for these experiments is very undesirable and can alter the results). The remaining ground connections don't carry switching current and so it's safe to connect them to the breadboards ground buss bar and jumper them through that to pin 10.

The lead length of decoupling capacitor C1 has been kept short to minimize its inductance. Decoupling allows the wires connecting +5V and ground from the bench power supply to the breadboard to be reasonably long (they were each 12" (30cm) in the prototype).

Parts List

Table A.1 lists the parts you'll need to perform the basic experiments. The following sections list specific part numbers and offer suggestions for substitutions. The cable is described in Appendix B.

Table A.1 Components required to perform the primary experiments.

Component	Quanity
74HCT240 integrated circuit	1
22Ω, 47Ω, 150Ω, 220Ω, 560Ω, 1kΩ, ¼ W fixed resistor	1 each
100Ω, ¼ W fixed resistor	4
100kΩ potentiometer	1
0.1µF capacitor	1
330pF capacitor	1
1N4148 diode	2
Cat5e cable	16.5 feet (6 meters)

You'll need to add one each of the following ¼ W fixed resistors if you wish to perform the supplemental exercises: 1Ω, 4.7Ω, 10Ω, 33Ω, 50Ω. The supplemental exercise also require a total of two 220Ω resistors, two BAT41 Schottky diodes, and a capacitor between 10 and 100pF.

Integrated Circuit

The 74HCT240 is a nice choice for this design because it's commonly available in leaded (through hole) packages. This lets experimenters build the circuit on solderless breadboards instead of requiring the manufacture of a printed circuit board. It also has enough individual inverters so that the entire circuit can be constructed from one integrated circuit. However, many other types of CMOS line driver or buffers are suitable. The important characteristics to look for are:

- High drive strength (output current drive strength of 6mA or greater).
- High impedance inputs (generally this means CMOS inputs).
- Low propagation time (since this usually means fast signal rise time).

Some suitable alternatives to the 74HCT240 include:

- CD40106
- 74HCT14
- 74HCT04
- 74AC04
- 74AC14

Note that if you do choose to use an integrated circuit other than the 74HCT240:

- To do all of the experiments you'll need a total of 8 inverters, so you'll need to use two integrated circuits if the replacement you choose has fewer than this.
- Ringing and other noise may become a problem if you substitute a part having a higher drive strength or faster rise time than does the 74HCT240 (which has a typical specification of 10ns), such as the 74AC series.
- You'll need a cable longer than what's described in Appendix B for the transmission line if the driver you use has a rise time longer than 10ns.
- If the alternative part you select is very fast you may need to add capacitance to slow down the ring oscillator and get it to operate over the range of frequencies need to reproduce the experiments (see the **Ring Oscillator** description in this appendix to find out how to do this).
- Be sure the driver doesn't have "open drain" or "open collector" type outputs.

Resistors

Fixed Resistors

Nearly any type of ¼ W leaded (through hole) resistors can be used in this circuit. Suitable resistor types include carbon film, metal film and (less commonly) carbon composition types.

- Note: The inductance of wire wound resistors makes them unsuitable and they must not be used in this application.

Thin or thick film surface mount resistors are quite suitable if the circuit is being built on a circuit board instead of a solderless breadboard. Power dissipation in the resistors is low during operation, so select the resistor body size based on your ability to solder them to the board. 0805 or 0603 sizes are easy to solder by hand, and their power ratings are more than adequate.

Potentiometer

Potentiometer *R1* can be of any type, provide it's not wire wound. Distributors sometimes classify these devices in their catalogs as "trimmers" rather than potentiometers or variable resistors.

The prototype used a single turn cermet trimmer, but any other non-inductive type can be substituted. In this application the improved resolution you'll get from a multi-turn potentiometer isn't necessary and isn't worth the higher cost.

Examples of suitable 100kΩ single turn trimmers include:

- Bourns Inc. part numbers: 3362P-1-104LF; 3386P-1-104LF; 3386F-1-104LF
- TE Connectivity part number: 1630480-4
- Copal Electronics part number: CT6EP104

Capacitors

Decoupling Capacitor *C1*

Decoupling capacitor *C1* should have a value from between 1nf and 100nf, and can be any type of ceramic dielectric provided it has a working voltage of 10V or higher. In their catalogs distributors sometimes classify these types of capacitors as "ceramic disk" (but be careful you don't get directed to large sized specialty high voltage capacitors), "radial disk" or "MLCC" capacitors. The type of capacitor used in the prototype can be seen in many of the photographs appearing throughout this manual (including Figure A.1).

Example suitable ceramic 100nF (0.1μF) capacitors include:

- Vishay Components Part number: K104K10X7RF5UH5
- Kemet: Part number: C315C104M5U5TA
- TDK: Part number: FK28X7R1H104K
- XICON: Part number: 140-50U5-104M-RC

Miniature aluminum electrolytic capacitors can be used provided the capacitance is at least 10μF and a working voltage of at least 10V.

Tantalum electrolytic capacitors of 1μF of higher also can be used, but these will cost considerably more than the ceramic types. Be sure the working voltage is no less than 10VDC.

Be sure to observe the proper polarity when using aluminum or tantalum capacitors.

Load Capacitor *CL*

Some experiments place additional capacitance at the transmission line's end to show the effect loading has on reflections. The prototype experiments specify a ceramic 330pF capacitor and CAT5e cable, but ceramic or mica capacitors in the 220pF to 560pf range perform well. Different capacitance may be necessary for other types of cable.

Example suitable ceramic 330pF capacitors include:

- Vishay Components Part number: F331K29Y5RN63J5R or S331K33Y5PR63K7R
- Kemet: Part number: C317C331J2G5TA
- TDK: Part number: FK18C0G1H331J
- AVX: Part number: SR151A331JAR

Diodes

Silicon small signal diodes are used as clamps in some of the experiments. The 1N4148 was used in the prototype, but any small signal diode will work, so don't be afraid to experiment with different types. Substitutions for the 1N4148 include the 1N914, 1N4448, and 1N457 (but don't mistake this for the 1N4571 Zener diode).

Schottky barrier diodes are used in the supplemental exercises. The BAT41 was used in the prototype, but suitable substitutions include the BAT42, 1N6263, 1N5819 and the SD101C.

Circuit Operation

As shown in Figure A.2, the test setup consists of a ring oscillator (inverters A, B and C and resistors *R1* and *R2*), a buffer (inverter D), and driver (inverters E and F and resistor *RS*).

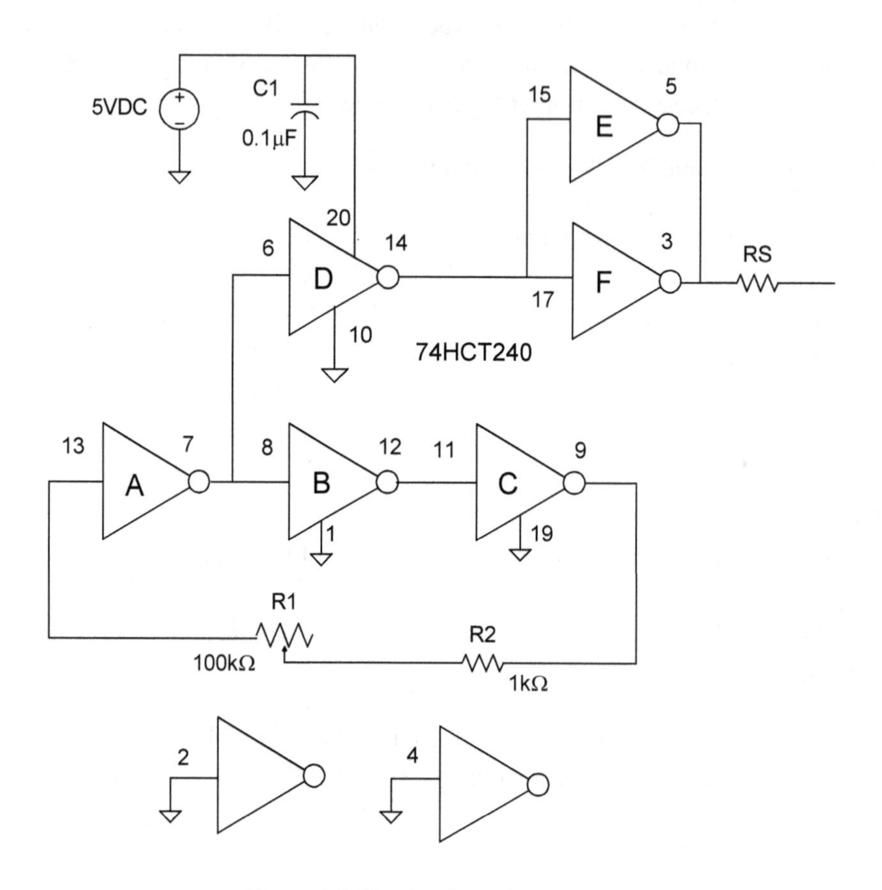

Figure A.2 Circuit schematic.

Ring Oscillator

Inverters A, B and C are connected as a ring oscillator. The sum of the delays of each inverter sets the delay through the entire ring, and so determines the period at which it oscillates. If we assume the propagation delay (*tpd*) of each inverter is identical (which usually isn't the case in practice, but is a good first order guess), a first order estimate is that the ring will oscillate with a period of

$t = 3 \times tpd$. Its frequency will be $f = \dfrac{1}{3 \times tpd}$. This ignores the delay caused by the input gate capacitance and parasitic capacitance.

Under worst case conditions *tpd* for a 74HCT240 is specified to be a maximum of 25ns. That makes the 3 stage ring oscillator operate with a maximum period of 75ns, which is a minimum frequency of 13MHz. Experience shows that many individual parts will oscillate faster than this, with periods under 50nS (20MHz) common when the circuit is built on a solderless breadboard.

This is faster than we need: For many of the experiments in this book the oscillator should have a period of 1µS or longer, but for other experiments we'd like the period to be under 100nS.

Resistors *R1* and *R2* allow the user to adjust the period of oscillation by adding an RC (resistor-capacitor) delay formed with the gate input capacitance on pin 13 (which is typically 15pF). This is enough capacitance to increase the period to about 1.5µS when *R1* is set to its maximum value. When *R1* is set to its minimum value the period is 100ns or less. Resistor *R2* limits the current through the wiper of potentiometer *R1* to levels safe enough for prolonged operation.

Small valued capacitors (in the 1 – 20pF range) can be connected from pin 13 to ground to add additional delay. This also makes the frequency of oscillation more repeatable because it's less dependent on the actual value of the gate input capacitance (which varies between individual integrated circuits). The timing uncertainty when using the input gate capacitance isn't a concern for the experiments described in this book because we'll be adjusting *R1* until we get the response we want from pins 3 and 5, regardless of whatever capacitance is actually present on pin 13.

Buffer and Driver

Inverter D buffers (isolates) the ring oscillator from the gate capacitance of inverters E and F (which are wired in parallel to increase their drive strength). This prevents the excess loading of gates E and F from affecting the ring oscillators period, but more importantly, it insures that pin 15 and 17 are driven with a sharp (fast) rise time signal. This helps to keep the rise time at pins 3 and 5 small, which is very desirable for the experiments in this book. By connecting pins 3 and 5 together we are tacitly assuming inverters E and F switch at the same time (that is, the two inverters have the same delays, and have the same rise and fall times). This is generally not the case: the two inverters switching behavior won't be precisely matched. The phase difference can cause high frequency noise spikes to appear on the power supply as the output of one driver is momentarily at a logic high value while the other is momentarily at a logic low (and vice versa). This briefly creates a low impedance path between the power and ground rails, which generates narrow glitches (spikes) on the power rail each time the driver switches. The glitching can be avoided by using two resistors for *RS* (one each connected to pin 3 and pin 5), but this technique complicates the impedance calculations you'll make in Chapters 2 and 3. The overlap noise was small in the prototype, so to keep things simple the circuit uses a single resistor for *RS*.

Resistor *RS* sets the output resistance of the circuit to a known value of our choosing. We need to know the value so the transmission line impedance can be calculated accurately. It's necessary to set the value because we don't know the actual output resistance of inverters E and F. Not only are they different for each individual integrated circuit, their outputs are transistors and not simple resistors. This makes their output resistance non-linear: Drawing different amounts of current from inverter pins 3 and 5 changes the apparent output resistance of inverters E and F. The output resistance also changes with temperature and voltage: The apparent resistance of the output falls as the integrated circuit gets colder, or when the value of the 5V supply is increased. Adding fixed resistor *RS* sidesteps these difficulties and makes the output resistance predictable.

Unused Gates

Many of the experiments use 6 out of the 8 available gates. The inputs of the unused gates are connected to ground as this prevents capacitive coupling that's present between the pins and on the die from causing the inputs to "float" to an intermediate voltage. Floating inputs of CMOS logic gates is not good engineering practice because it can result in the integrated circuit generating high frequency noise and drawing larger than expected current.

Appendix B. Selecting and Preparing the Cable

The experiments in this book were performed with solid core CAT5e cable. Any multi-conductor cable will work, but it must have at least two wires. You'll need a cable with more than two wires if you want to perform the crosstalk experiments. If you use a different type of cable the results you get for impedance (and to a lesser extent, delay) won't match the example results, but the concepts will be the same and you'll still be able to draw the correct conclusions.

Why CAT5e?

CAT5e was chosen because it's inexpensive, widely available, has lots of wires and the impedance is uniform even for long lengths. The wire size (24AWG) works well with solderless breadboards. Stranded core wire also works, but it's difficult to get a reliable connection when it's used with solderless breadboards. If you do choose to use stranded core wire (of any type), solder a short stub of solid wire to the ends of the stranded wire, and plug these rather than the loose strands into the breadboard. Number 24AWG works well for the stubs, or you can use the discarded leads from a 1/4W resistor or a small leaded capacitor.

CAT5e cable has 8 total wires, but most of the experiments can be done with a cable of only 4 wires. In fact, the basic experiments can be done with only two wires (such as speaker wire or the common "zip cord" used to wire lamps or small mains operated appliances), but you won't be able to measure the effects of crosstalk and other similar kinds of coupling. For this reason 4 or more wires is preferred.

Other Wire Choices

CAT6 and CAT7 wire can be used, but these have lower crosstalk specifications than CAT5e. Although this characteristic improves high performance signaling in systems, lower crosstalk for the cable means smaller crosstalk voltages will be measured in the Chapter 7 experiments. This is one of the few times where more crosstalk is better than less, so to avoid difficulties in making the crosstalk measurements CAT5e was chosen rather than CAT6 or 7 cables.

The multi-conductor "alarm wire" used to connect sensors to an alarm control panel is inexpensive and can be used in place of the CAT5e wire. It usually consists of 3 to 4 (or sometimes more) solid core wires, but the impedance isn't controlled and so the results you obtain won't be as clean looking as you'll get with the CAT5e wire. Be sure the wire you select isn't too thick to fit into the solderless breadboard sockets. If need be, attach a short stub of smaller diameter wire, as is described above for stranded wire.

Thin gauge two conductor solid core speaker wire can be used for the more basic impedance and delay measurements, but you won't be able to perform the crosstalk experiments. Although any wire gauge can be used, you'll find wires in the 22 to 24 AWG range the easiest to use with a solderless breadboard.

Using Coax Cable

Coax cable can be used to perform the basic experiments, but you won't be able to make crosstalk or coupling measurements. The miniature 50Ω RG316 is easy to use and is widely available in spools. You'll need at least 13 feet (4 meters), but it may need to be longer than this if the integrated circuit you chose for the driver has slow rise times, or if the bandwidth of your oscilloscope is less than 200MHz. Shorter lengths may work for you, but success will depend on how well you control ringing and reflections caused by your oscilloscope probe ground connections. These things are described in Appendix C.

Other 50Ω cables such RG58 can also be used, but their larger diameter makes them more difficult to work with. You'll need at least 13 feet (4m).

A less expensive alternative to RG316 or RG58 is the 75Ω coaxial cable used to connect televisions and other video equipment. It comes in many different styles, but the foam core with a solid conductor version is often the least expensive because of its wide use.

Making Your Own Cable

Cables are better than individual strands placed near each other and secured with tape or wire ties because the wire to wire spacing is more consistent in a cable than for loose wires (even when they are bundled closely together). This consistency is critical for repeatable experimental measurements, and it produces cleaner waveforms.

However, you can make your own multi-conductor cable if you happen to have enough wire on hand. For best results form a cable by twisting the 4 or more individual wires around each other. Stranded or solid core wire of any diameter can be used, but AWG22 to 24 gauge wires are especially convenient. An easy way to make a cable having reasonably repeatable electrical characteristics is to secure one end of all the wires, then place the other end of all the wires in the chuck of an electric drill and use it to twist the wires together. Besides being faster than twisting them by hand, the drill produces a uniform number of twists per inch. Roughly 3 or 4 twists per inch work well, but for these experiments the number of twists isn't critical provided the cable retains its shape once its ends are removed from the drill. A uniform twisting is desirable as it'll give cleaner, more repeatable experimental measurements.

How Long Must the Cable Be?

Although the precise length isn't critical, it'll need to be long enough to add an amount of delay no less than the rise time of the signal the driver puts out. The delay must also be long enough to be clearly seen on your oscilloscope. For the 74HCT240 type driver used in the prototype experiments, a length of cable giving a delay of about 25ns works well with a 200 MHz bandwidth oscilloscopes. The cable can be slightly shorter if your oscilloscope has a higher bandwidth. Longer lengths are easier to measure on the oscilloscope but are more difficult to physically manage. Very long lengths should be avoided as wire losses will become high enough to distort the measurement results.

The suggested lengths when using an HCT240 driver and a 200MHz oscilloscope are:

- CAT5e, CAT6 or CAT7: 16.5 feet (5 meters)
- Alarm wire: 20 feet (6 meters)
- Zip cord or speaker wire: 20 feet (6 meters)
- Coaxial cable: 13feet (4 meters)

Double these lengths if you are using a 100MHz bandwidth oscilloscope.

Preparing a CAT5e Cable

Cut a length of CAT5e cable 16.5 feet (5 meters) long, and remove 3 ½ to 5 inches (9 to 13cm) of the outer insulation from both ends. As is shown at the right in Figure B.1 the manufacturer has twisted the wires into 4 pairs. This is done to control impedance and to reduce coupling (crosstalk) between pairs.

Color identifies individual wire pairs; one wire in the pair has a solid color (orange, green, blue or brown), while its companion is white with a thin, faint stripe of that same color.

Figure B.1 CAT5e cable.

To make the wire easer to work with it may be coiled into a loop and secured with cable ties as is shown in Figure B.2. Tape can also be used. Because of the twisted pairs, coiling the wire won't alter the experimental results provided the loop isn't so small that individual wires become bent. The coil in the Figure has a diameter of 5 ½ inches (14cm).

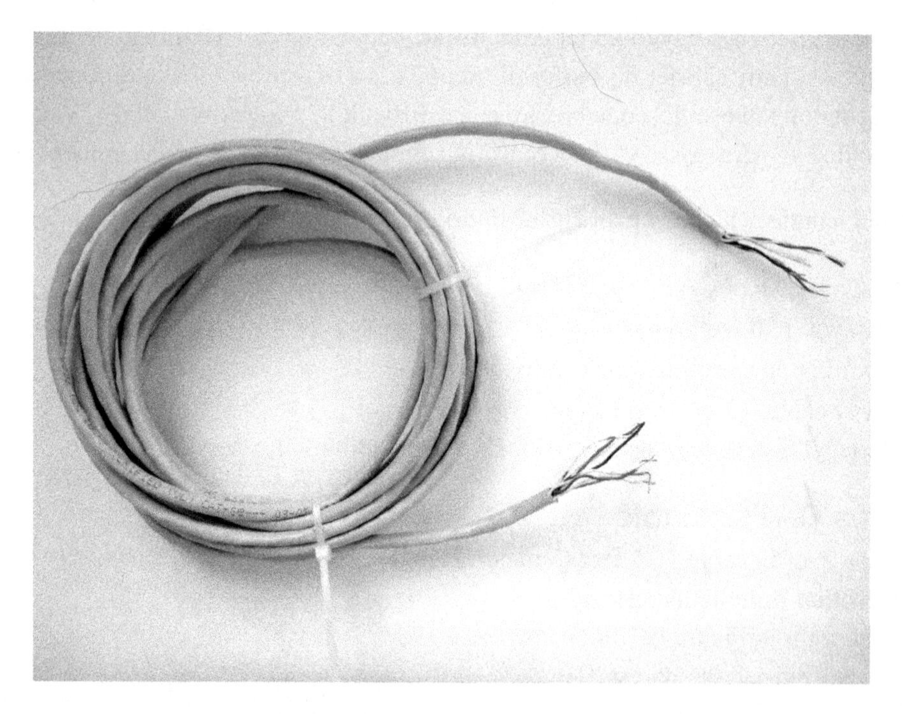

Figure B.2 Form cable into a loose coil. Secure with cable ties (shown), or tape.

Complete the cable preparation by spreading out the pairs as is shown in Figure B.1. Then untwist the blue pair and remove about ¼ inch (0.6cm) of their insulation.

Appendix C. Oscilloscope Probing Techniques[4]

You can use a wire or piece of common coax to directly connect an oscilloscope to the circuit you're testing, but because they have higher impedance, you'll generally use the probe that came with your oscilloscope. Most probes are "10X" (they divide the signal by 10), but some can be switched between 1X and 10X. The impedance and bandwidth will be highest when it's set to the 10X position. To get the best measurement results you'll need to understand a few critical things about the electrical characteristics of your probe.

Oscilloscope Probes

Figure C.1 shows a popular 200MHz 10X probe. You can see the probe requires two connections to your circuit: The probe tip itself and a ground connection. The ground can be connected with either a wire attached to an insulated alligator clip, or with short springs that are attached directly to a ground ring on the probes body. Keeping the inductance of the ground connection low will improve the accuracy of your measurements. As you'll see, the ground springs are electrically much better than using the long ground wire, but the wire version is provided because its length is more convenient. The springs can be purchased from the probe manufacturer, or you can make your own by tightly winding 3 or 4 turns of #26 or #24 sized solid copper wire around the probes ground ring. Cut the end to whatever length is necessary in your application, but keep it as short as you can. Don't use stranded wire.

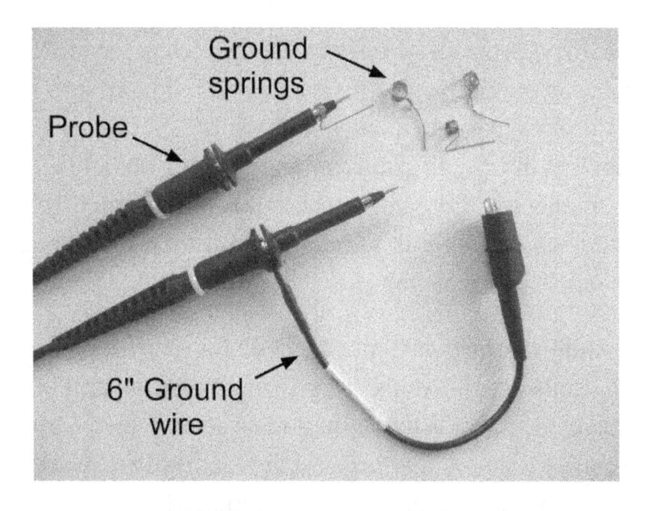

Figure C.1 200MHz oscilloscope probe with ground wire and ground springs.

[4] See Bibliography for references

Probe Circuit Model

A well designed oscilloscope probe presents a minimal parasitic load to the circuit you're testing, but even the best probes will add capacitance and inductance to the node it's measuring, and will siphon away a small amount of current from the circuit under test to the oscilloscopes vertical amplifier. A simplified drawing of a typical 10X probe and the connection to the oscilloscope appears below, in Figure C.2.

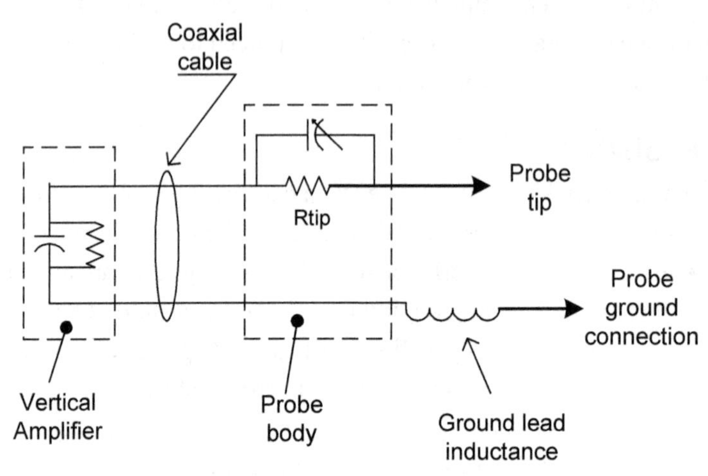

Figure C.2 Simplified oscilloscope probe electrical model.

The probes input resistance is mostly set by resistor *Rtip*, which is located inside the probes body. The compensation capacitor in parallel with *Rtip* is adjusted by the user to make the frequency response of the probe uniform over its rated bandwidth. This capacitance along with other parasitic capacitances presents a load to the circuit you're probing, and is one of the reasons why the probes impedance changes with frequency. More important to this discussion is that the inductance of the ground lead forms a parallel resonant circuit with *Rtip* and its capacitance, and the resistance and capacitance of the oscilloscopes vertical amplifier.

To see this we must first know that *Rtip* is typically 9MΩ for many high-impedance 10X probes designed to connect to an oscilloscope having an1MΩ input resistance. This gives the desired 10X scale factor, but it also means that *Rtip* is the highest resistance in the circuit. This lets us take a huge leap and simplify things by saying that because the input resistance of the vertical amplifier is small compared to *Rtip* it can be ignored and so can be replaced with a short circuit. This analysis isn't fully correct, but our concern here is to understand how inductance in the probes ground lead can cause resonance and not to get sidetracked by a lengthy but more technically correct analysis which would lead us to the same conclusion anyway.

We'll model the circuit you're measuring as a load resistor (*Rload*) in series with a perfect voltage source *Vload* (one having no impedance).We'll make one other assumption and say that the impedance of your circuit is much lower than the value of *Rtip*. This must be true; otherwise this particular probe would excessively load down your circuit. You would switch to a higher impedance probe (such as a FET probe, or a high impedance passive probe, one having *Rtip* values greater than 9MΩ) when this happens. Although *Rload* is small compared to *Rtip*, its resistance isn't zero. We'll return to this important observation shortly.

With these assumptions in place we can now see how the ground lead inductance is in parallel with *Rtip* and the compensation capacitor. Figure C.3 at the left shows the circuit model. If *Rload* was zero *Lgnd* would be directly in parallel with *Rtip* and *Ccomp*, creating a very high-Q parallel RLC circuit. This circuit can produce resonance voltages much higher than the value of *Vload*. This voltage appears to the oscilloscope as an error voltage that adds to the actual signal you're measuring.

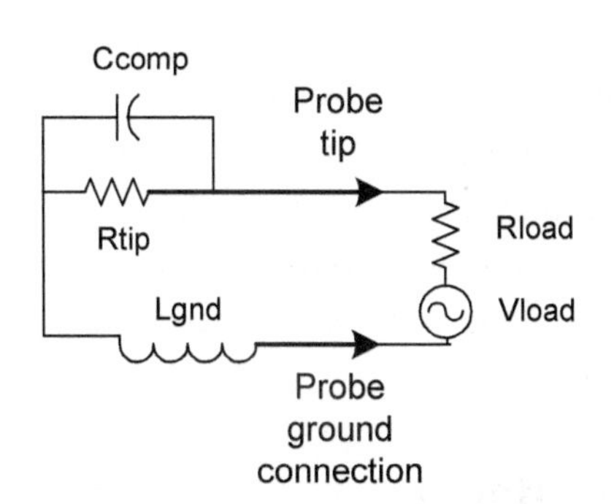

Figure C.3 Ground inductance is in parallel with probe capacitance and resistance, forming a parallel RLC tank circuit.

Lowering the circuit Q will reduce the magnitude of the ringing (the error voltage). This can be done by lowering the inductance, or by adding resistance in series with it to lower the Q.

Reducing Ringing By Lowering Ground Inductance

From field theory we know that the inductance of a wire depends on how long it is, and how far away it is from its "return path". Chapter 2 briefly discussed return paths. Here all we need to know is that a long wire that's far away from its return will have higher inductance than a short wire that's close to the return such as a ground plane. In practical terms this means when probing circuits always use the shortest ground lead you can, and try to keep the ground wire as close to the probe body as possible. Often this isn't practical, but this explains why the ground inductance is so much lower when using the ground spring instead of the wire and alligator clip: The spring is quite a bit shorter and is very close to the probe body (the tip in particular). The inductance of the wire and clip arraignment is typically in the 150nH range; the spring is usually under 20nH. These differences are evident in Figure C.4 (which we'll discuss shortly).

Manufactures of high quality probes offer small sockets that you can solder onto your board. The probe is plugged into the socket and its outer shell makes contact with the probes ground ring. The probe tip connects to a short lead that's soldered to the signal you're testing. These sockets provide an excellent low inductance connection between the probe ground and the ground system of the board you're probing, but on high density circuit boards it's not always possible to find the space necessary to mount them.

Reducing Ringing by Adding Resistance

Lowering the circuit Q by adding resistance is one of the more surprising results of the circuit analysis we've just performed. This brings up two critical observations:

1. Any artificial ringing you see on the oscilloscope screen depends on the value of *Rload*. In fact, your oscilloscope will display less probe ringing when the circuits you're probing have high resistance. This means false ringing will be higher for low resistance circuits such as power supply circuits or very low output impedance logic gates. False ringing will be less when probing high resistance logic circuits or opamp inputs, for example.
2. Adding a resistor in series with *Lgnd* will reduce the ringing. If done properly this won't affect the accuracy of the measured waveform, but it will reduce or eliminate the false ringing. In practice, this is done by adding a resistor in series with the ground lead, usually by attaching a resistor to the alligator clip at the end of the ground wire. Resistance values from 33Ω to as much as 200Ω can be used.

Effects of Reducing Inductance and Adding Resistance

The effects of adding resistance in the ground lead and reducing inductance by keeping the ground lead short is shown in Figure C.4. The signal should rise smoothly and level off at 3.3V, and there should be no ringing or overshooting present, but we can see how inductance alters the waveform displayed by the oscilloscope.

The graph in the upper left hand corner is the response when the ground lead is 20" (51cm) long. The figure to its right shows the same signal when the ground wire has been shortened to 6" (15cm). Both graphs show ringing, overshooting and ring back, none of which is actually present in the signal.

The graph in the lower left shows the improvement when a 51Ω resistor is placed in series with the 6" ground lead. The overshoot has been reduced greatly, and the ring back has been all but eliminated. Although not shown, increasing the resistance to 100Ω or 200Ω further improves the waveform, but making the resistor too large distorts the signal and introduces rounding which is not present in the actual signal.

The graph in the lower right is the oscilloscope display when the ground spring is used rather than the ground wire.

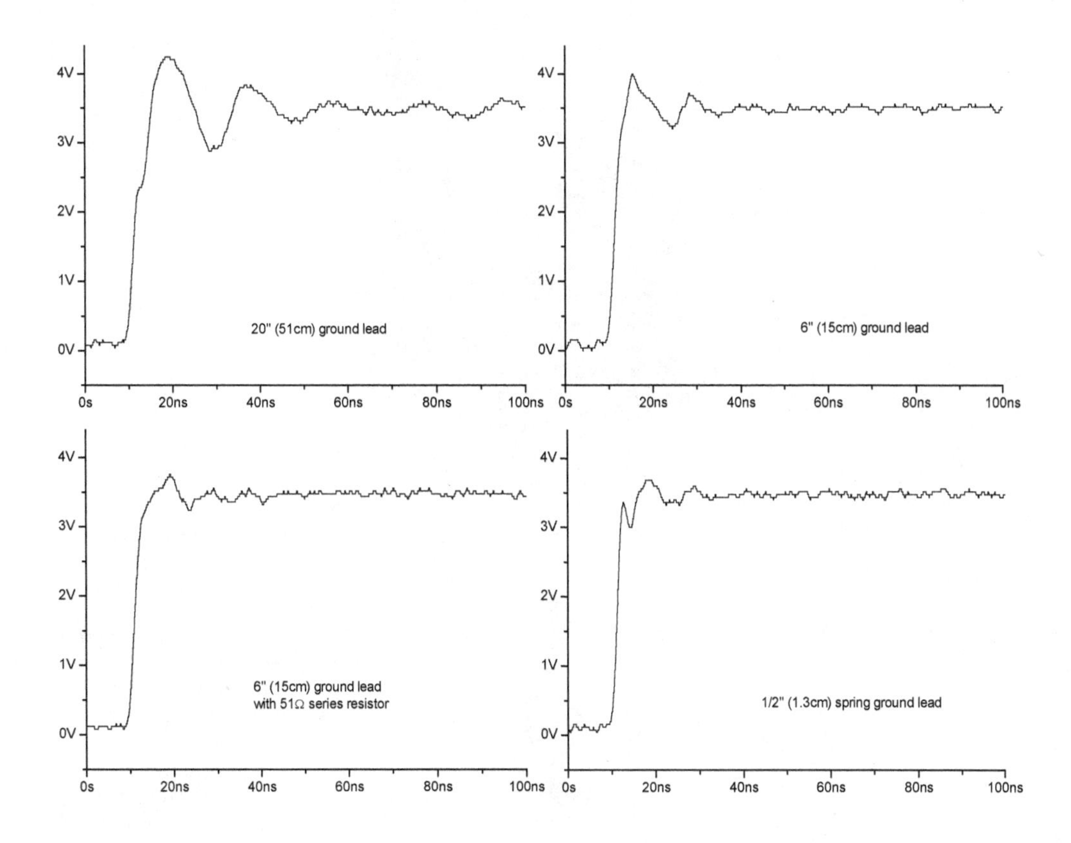

Figure C.4 Effects of ground inductance and how ringing can be improved by adding resistance.

Using a Ground Spring with a Solderless Breadboard

Many of the measurements presented throughout this manual were made with the ground spring used as the probe ground. The photograph in Figure C.5 shows the setup. The ground is short and the distance between the probe tip and ground is small, reducing the size of the return path, and so, the inductance.

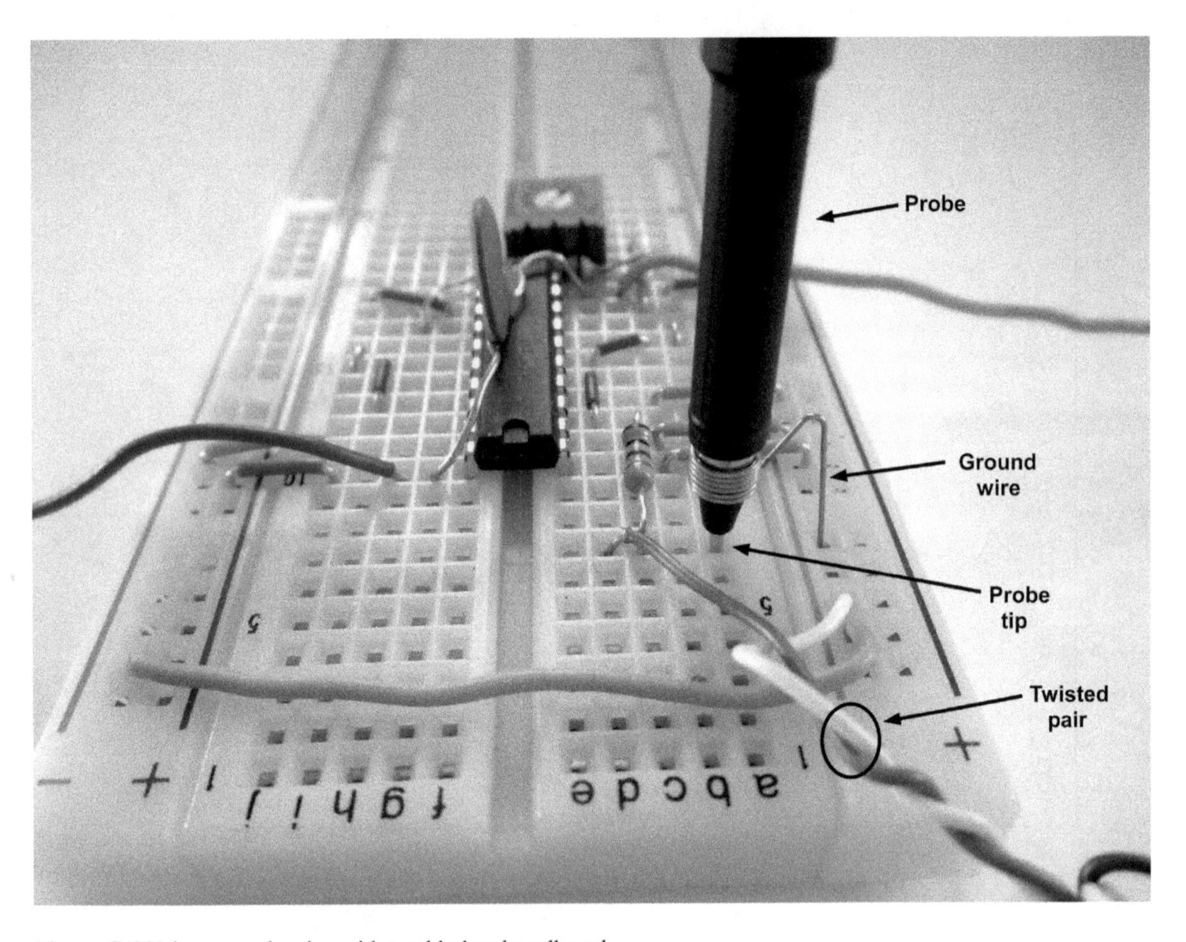

Figure C.5 Using ground spring with a solderless breadboard.

Oscilloscope Bandwidth and Accuracy

It's natural to think the oscilloscopes bandwidth specification tells you the highest frequency the oscilloscope can faithfully display, but this isn't so. Actually, the oscilloscope and its probe are a low pass filter that will cause the display to misrepresent the amplitude for frequencies well below the bandwidth specification. In fact, the bandwidth is the "3dB cutoff frequency" of the oscilloscopes vertical amplifier when it's connected to the particular probe identified by the manufacturer. Using a probe from a different manufacturer (or a lower or higher bandwidth probe from the same manufacturer) can change the oscilloscopes frequency response and give you unpredictable results.

Frequency Response

So, what is the "3dB cutoff frequency"? From filter theory we know it's that frequency where the output amplitude has been reduced by 3dB from the input amplitude. A loss of 3dB means the amplitude will only be 0.707X the original value. For an oscilloscope this means a signal measured at that frequency will be displayed on the screen with a value of 70.7% of the actual value. Higher frequencies will be attenuated more and so will displayed with lower values.

The theoretical response for a 200MHz bandwidth oscilloscope when connected to the correct 200MHz probe is shown below, in Figure C.6.

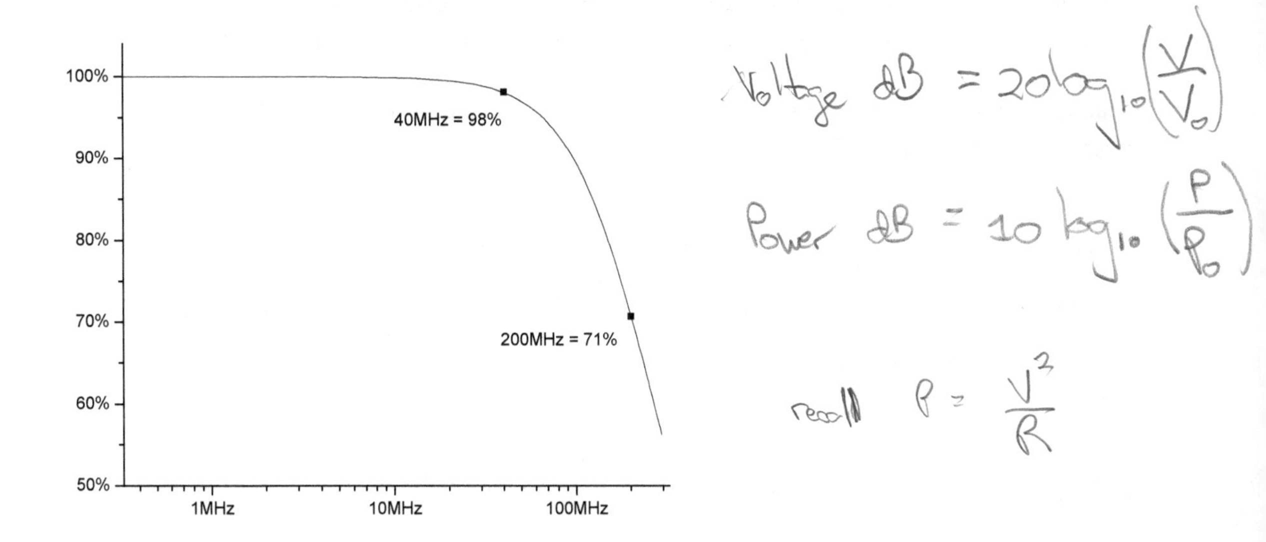

Figure C.6 Theoretical frequency response of a 200MHz oscilloscope probe.

You can see that for a 200MHz sine wave the oscilloscope displays a amplitude of only 71% of the actual signal value in the graph (the actual value of 70.7% has been rounded to 71%). For example, a 1V peak-to-peak 200MHz sine wave would be displayed as 0.71V peak-to-peak.

A pulse is made up of many sine wave components (its harmonics), and the oscilloscopes frequency response will distort the shape of the pulse as each frequency is attenuated by a different amount. We'll return to this shortly.

The figure also shows that the displayed amplitude error is small for low frequencies, but grows quickly for frequencies above about 40MHz. This brings us to the often quoted rule of thumb:

Divide the oscilloscopes rated bandwidth by 5 to estimate the highest frequency it can accurately measure.

This is verified in the figure: The oscilloscope will display 98% of the signal's actual amplitude at 40MHz (which is 1/5th of the oscilloscopes 200MHz rating).

Roll of Ground Lead Inductance in Changing the Frequency Response

The previous figure assumes the probe and oscilloscope is a first-order low pass filter (in fact, an RC filter), but inductance in the ground lead actually makes it a second-order filter. Second-order filters have resonances, and their frequency response can be very non-linear. For this reason manufacturers assume the probe is connected to ground near its tip through a low inductance path when they specify the bandwidth since this gives the most predictable (and most desirable) results.

The actual measured response of the 200MHz oscilloscope and probe used throughout this manual appears below, in Figure C.7.

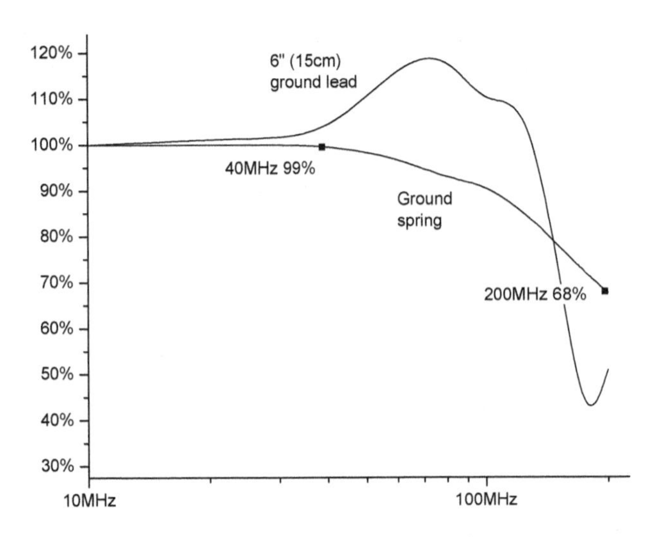

Figure C.7 Measured response of the 200MHz probe used in this manual. Using the ground spring gives a response very close to the theoretical.

From the previous discussion we'd expect the curve to have the shape shown in Figure C.6, and the "Ground spring" curve does closely resemble the ideal shown in that figure. However, the "6 inch ground lead" curve is quite a bit different. It shows gain for signals in the 40MHz to 100MHz range, than shows a very rapid fall off in gain for higher frequencies. This means high frequency signals will be distorted. For instance, when using the probe with a 6" ground lead a 1V peak-to-peak, 80MHz sine wave would display as 1.2V peak-to-peak. A 1V, 20MHz sine wave would display properly. In the time domain this frequency distortion will show up as misshaped pulses and ringing which is not present in the actual waveform.

Observing Pulses

The harmonics making up a pulse must be received with the proper amplitude and phase, otherwise the pulse will appear distorted. Pulses having sharp rise and fall times contain more high frequency harmonics than long rise and fall time pulses. This makes it necessary for the oscilloscope bandwidth to be higher than just the frequency of the pulse train you'll be measuring, otherwise the displayed pulse won't match the actual one very closely. For instance, you'll need a higher bandwidth oscilloscope to properly display the shape and amplitude of a 10MHz square wave that has a 1ns rise time than you will if the 10MHz square wave has a 10ns rise time. Filter theory can be used to estimate the highest frequency present in a pulse based on its rise time (by "rise time" we mean either the rise or the fall time, whichever is smallest). From this we can then estimate the oscilloscope bandwidth necessary to properly observe the pulse.

The filters bandwidth (*BW*, the highest frequency of interest) is determined with equation (C.1) when the pulse rise time (*tr*) is known:

$$BW = \frac{0.35}{tr} \tag{C.1}$$

The 0.35 constant is an approximation. The actual value depends on how smooth the rise time is, and how much ringing is present. Values closer to 0.5 are used when the pulse contains lots of high frequency ringing or other noise. These details don't concern us here, because we'll be using the equation to estimate the bandwidth required of your oscilloscope so it accurately makes pulse measurements. We're not using it to predict the actual response.

The 74HCT240 integrated circuit was chosen for the experiments in this manual in part because its rise time (typically 10ns) is slow by modern standards. This allows us to get the proper results even when using a relatively low bandwidth oscilloscope.

By using the equation we find that the pulses will have useful energy for frequencies as high as $BW = \frac{0.35}{tr} = \frac{0.35}{10ns} = 35MHz$. This is as true when R1 is adjusted so the ring oscillator is producing low frequency pulses (say, 1MHz pulses) as it is when the ring oscillator is producing 10MHz pulses.

From the previous discussions we know that the oscilloscope and probe frequency response isn't "flat" (consistent) unless the oscilloscope bandwidth is 5 times greater than the highest frequency we're trying to measure. Since we need to measure frequencies as high as 35MHz, this means we need an oscilloscope having a bandwidth of about 35MHz x 5 = 175MHz.

If your oscilloscope doesn't have this high of a bandwidth you can still perform the experiments. The waveforms you'll see won't closely match the ones presented in this manual, and you're oscilloscope may miss some of the high frequency reflections. But you should still see the effects terminations have on reflections and the experiments illustrating impedance, delay and crosstalk will still work for you (although you'll lose significant detail if your oscilloscope bandwidth is less than roughly100MHz). One advantage of using low bandwidth equipment is that grounding your scope probe isn't as important as it is for high bandwidth oscilloscopes. The high frequency ringing introduced by long ground leads still disturbs the circuit under test, but the oscilloscope isn't fast enough to capture it.

Summary of Measurement Tips

- Always strive to use the lowest inductance ground connection with your oscilloscope probe. Add local test grounds to your circuit board if necessary to make it possible to ground your probe with a short lead.
- Spring clips or special oscilloscope probe sockets have the lowest inductance and so will give the best results.
- You can make your own "spring clip" by tightly winding a few turns of #26 or #24 wire around the probes ground ring.
- Connecting the oscilloscope probe to the circuit under test with a poor, high inductance ground causes the oscilloscope to display false ringing, and results in incorrect amplitude measurements.
- False ringing and overshooting can be eliminated by adding small amount of resistance (usually less than 200Ω) in series with the alligator clip ground wire that comes with your probe.
- A good way to determine if the ringing you see is really in your circuit or if it's an artifact of ground inductance is to add a 100Ω resistor in series with the ground lead. If the ringing goes away, or is reduced in amplitude, then ground inductance is the most likely cause.
- To accurately measure the amplitude of sine waves the bandwidth of your oscilloscope should be at least 5X greater than the sine wave frequency.
- You can estimate the highest harmonic frequency present in a pulse train with equation (C.1). Multiply that value by 5 to determine the oscilloscope bandwidth you'll require.

Bibliography

Over the years many textbooks and journal articles have been written regarding signal integrity. In my view, the best of these were written in the period from the late 1980's through the middle of the 2000's. These works cover the basics and usually don't assume the reader has previous knowledge or background in RF, electromagnetics or high-speed signaling. More recent works are valuable in that they provide guidelines and rules for solving contemporary signal integrity problems. Engineers and technicians whose work involves solving modern signal integrity problems should have at least one of these books on their bookshelf.

However, when learning the fundamentals the older books are superior to more recent works that present overly simplistic rules of thumb or metrics. The older works provide the foundation necessary to understand the proper use of the rules and suggestions presented in more recent books.

This bibliography presents texts and journal articles that may not be familiar to the newly graduated engineer, or to an experienced engineer that has been exposed to signal integrity through one of the more recent texts. They provide important background material and will help the practitioner develop the strong foundation necessary to solve difficult and unordinary signaling problems.

Electromagnetics

The books in this grouping discuss the physics of electromagnetics. All of them involve mathematics, including advanced mathematics in some cases. This especially applies to the books by Ramo and Magnuson, which take a deeply theoretical approach to field theory.

This contrasts with Sibley and Hammond's books, which are entry level and use a minimum of mathematics.

Miner's book remains my favorite as an all-purpose, well-round electromagnetics/field theory text.

My first book on signal integrity (*High-Speed Circuit Board Signal Integrity*) is a hybrid electromagnetics and signal integrity text. It uses simple mathematics to explain the physics of high-speed signaling at the circuit board level. It's been particularly popular with custom chip designers, ASIC designers and test engineers wishing to understand how resistance, inductance and capacitance is formed on a circuit board, and the electrical trade-offs necessary when signaling at high speeds over circuit board traces. The differences between traces buried deep inside a circuit board stackup vs. those routed on the surface are discussed in detail. Crosstalk and signal losses are presented in depth.

- Brown, R.G., et-al., *Lines, Waves and Antennas, 2ed edition*, New York: John Wiley & Sons, 1973
- Hammond, P., *Electromagnetism for Enginee*rs, 3ed edition, Oxford: Pergamon Press, 1986
- Magnusson, P.C., et-al., *Transmission Lines and Wave Propagation*, 3ed edition, Boca Raton: CRC Press, 1992
- Miner, Gayle, *Lines and Electromagnetic Fields for Engineers*, New York: Oxford University Press, 1996
- Ramo, S., et-al., *Fields and Waves in Communication Electronics*, 3ed edition, New York: John Wiley & Sons, 1994
- Sibley, M., *Introduction to Electromagnetism*, London: Arnold Press, 1996
- Thierauf, S.C., *High-Speed Circuit Board Signal Integrity*, Norwood: Artech House, 2004

Transmission Lines, Terminations and Reflections

The listed books by Matick and Bakoglu are classics and in many ways paved the way for more recent signal integrity texts that are less theoretical. Bakoglu's book was written for the CMOS chip designer, but his description of transmission lines and losses is clear, concise and approachable. Those chapters are still valid today, and are useful to anyone interested in understanding transmission line theory. Matick's book is more theoretical and is very through. The serious practitioner should have a copy on their bookshelves.

The book by Hart is a succinct presentation of transmission lines, terminations and crosstalk. Mathematics is kept to a minimum. This is an excellent book for the beginner.

The application notes by National Semiconductor and Motorola are standards and are easily read. They provide concise design information, and explain transmission lines and reflections without much mathematics. They are worth finding on the web and adding to your library.

Rosenstark's book nicely describes transmission lines and their reflections (including the way in which loading along a transmission line changes the impedance and time of flight), but his focus is on ECL ('emitter coupled logic', which uses bipolar transistors) rather than the CMOS logic now in common use. Nevertheless, his book is recommended as an easy entry into the subject of transmission lines.

A more advance work but one that's still suitable for the beginner is the undergraduate text written by Sinnema. His book discusses transmission line theory, including propagation, impedance and loss. It's fundamentally an electromagnetics and propagation text, but the mathematics used is simple and quite suitable for an associate degreed technician or the degreed undergraduate engineer. This book does not discuss distributed loads or other practical signal integrity concerns (those topics are covered in Rosenstark's text).

Even today W.C. Johnson's book is by far the best transmission line text available in my view. It's highly recommended to those wishing to obtain more than a casual understanding of transmission lines. The focus is long wire transmission lines such as telephone and power distribution (mains) transmission lines, but he does discuss RF circuits. Johnson's discussion of reflections and losses is unmatched. The book is mathematical in nature but not overly so and is an excellent introduction to advanced transmission line theory. Although long out of print used copies do become available from time to time.

My second book on signal integrity (*Understanding Signal Integrity*) is an entry level text that describes transmission lines, signal loss, crosstalk, reflections and terminations. This lab manual grew out of the requests from educators for a laboratory guide to accompany that book. Reflections (including bounce diagrams) and routing topologies are discussed in detail, and it contains student problems/exercises. It's an excellent introduction to my first signal integrity text (*High-Speed Circuit Board Signal Integrity*).

- Bakoglu, H. B., *Circuits, Interconnections and Packaging for VLSI*, Reading: Addison-Wesley Publishing Company, 1990
- Dally, W.J., Poulton, J. W., *Digital Systems Engineering*, Cambridge: Cambridge University Press, 1998
- Hall, et al., *High-Speed Digital System Design*, New York: John Wiley & Sons, 2000
- Hart, B.L., *Digital Signal Transmission Line Circuit Technology*, London: Van Nostrand Reinhold (UK), 1988
- Johnson, W. C., *Transmission Lines and Networks*, New York: McGraw Hill, 1950
- Matick, R.E, *Transmission Lines for Digital and Communication Networks*, New York: IEEE Press, 1969
- Montrose, M., *EMC and the Printed Circuit Board*, New York, IEEE Press, 1999
- Motorola Semiconductor, *Transmission Line Effects in PCB Applications*, Application Note AN1051
- National Semiconductor, *Transmission-Line Effects Influence High-Speed CMOS*, Application Note 393

- Royle, David, *Rules tell whether interconnections act like transmission lines (Parts 1,2,3)*, EDN, June 23, 1988, pp 131 – 136; 143 – 148; 155 – 160
- Rosenstark, S., *Transmission Lines in Computer Engineering*, New York: McGraw Hill, Inc., 1994
- Sinnema, W., *Electronic Transmission Line Technology, 2ed edition*, New Jersey: Prentice Hall, 1988
- Thierauf, S.C., *Understanding Signal Integrity*, Norwood: Artech House, 2011
- True, Kenneth, *Reflections: Computations and Waveforms*, National Semiconductor Application Note 807, March 1992

Crosstalk

The previously listed books by Bakoglu, Hall, Hart, Montrose, and my two books discuss crosstalk without going into a great deal of mathematics. The beginner wishing to understand crosstalk should have one or more of these texts for reference. These are in stark contrast with Paul's book, which is very complete but isn't for the beginner. Young's text treats crosstalk mathematically and is valuable for the advanced practitioner.

DeFalco's paper is the classic source that describes crosstalk simply and is worth obtaining by the beginner or advanced practitioner.

- Catt, I., "Crosstalk (Noise") in Digital Systems," *IEEE Trans. On Electronic Computers*, EC-16, No. 6, December 1967, pp. 743-763
- DeFalco, J., "Reflections and Crosstalk in Logic Interconnections," *IEEE Spectrum*, July 1970, pp. 44-50
- Feller, Al, et al., "Crosstalk and reflections in High-Speed Digital Systems," *Proceedings of the Fall Joint Computer Conference*, Washington, D.C., December 1965, pp. 511-515
- Paul, C. R., *Analysis of Multiconductor Transmission Lines*, New York: John Wiley & Sons, 1994
- Walker, C.S., *Capacitance, Inductance and Crosstalk Analysis*, Norwood: Artech House, 1990
- Young, B., *Digital Signal Integrity*, New Jersey: Prentice Hall, 2001

ESD and CMOS I/O circuits

The following references give a good description of why and how ESD clamping devices are constructed on CMOS integrated circuits. The books by Chandrakasan and Dabral detail the operation and describe the failure mechanisms that can occur when the I/O circuits are stressed. Application notes 77 and 140 are more general in nature and are helpful in understanding the basics.

- Buurma, G., *CMOS Schmitt Trigger – A Uniquely Versatile Design Component*, National Semiconductor Application Note 140
- Calebotta, S., *CMOS, the Ideal Logic Family*, National Semiconductor Application Note 77
- Chandrakasan, A., et.-al., *Design of High-Performance Microprocessor Circuits*, New Jersey: IEEE Press, 2001
- Dabral, S., et.-al., *Basic ESD and I/O Design*, New York: John Wiley & Sons, 1998
- Kulkarni, V., *Electrostatic Discharge Prevention – Input Protection Circuits and Handling Guide for CMOS Devices*, National Semiconductor Application Note 248

Oscilloscope Probing

Doug Smith's book is the classic text describing proper oscilloscope probing techniques when probing below a few 100 MHz. It was my original source for many of the ideas appearing in Appendix C. Although those making measurements on high-performance interconnect such as Gbs serial signaling will find it somewhat outdated, it's nevertheless highly recommended, especially for those just learning to use an oscilloscope.

- Montrose, M., Nakauchi, E., *Testing for EMC Compliance*, New Jersey: IEEE Press, 2004
- Smith, D., *High Frequency Measurements and Noise in Electronic Circuits*, New York: Van Nostrand Reinhold, 1993
- Tektronix, *ABCs of Probes Primer*, Tektronix, Document number 60W-6053-8, 2004

CPSIA information can be obtained
at www.ICGtesting.com
Printed in the USA
FSHW012023240619
59403FS

9 781500 480516